SpringerBriefs in Animal Sciences

More information about this series at http://www.springer.com/series/10153

Ivica Králová-Hromadová
Ludmila Juhásová · Eva Bazsalovicsová

The Giant Liver Fluke, *Fascioloides magna*: Past, Present and Future Research

Ivica Králová-Hromadová
Institute of Parasitology
Slovak Academy of Sciences
Košice
Slovakia

Eva Bazsalovicsová
Institute of Parasitology
Slovak Academy of Sciences
Košice
Slovakia

Ludmila Juhásová
Institute of Parasitology
Slovak Academy of Sciences
Košice
Slovakia

ISSN 2211-7504 ISSN 2211-7512 (electronic)
SpringerBriefs in Animal Sciences
ISBN 978-3-319-29506-0 ISBN 978-3-319-29508-4 (eBook)
DOI 10.1007/978-3-319-29508-4

Library of Congress Control Number: 2016930439

Printed on acid-free paper

This Springer imprint is published by SpringerNature
The registered company is Springer International Publishing AG Switzerland

This publication is dedicated to three special people to whom we would like to express our most sincerest gratitude for their intensive help, long-standing support, and significant contribution to our research of Fascioloides magna*, which inspired us to write the publication.*

Marta Špakulová is the senior researcher at the Institute of Parasitology, Slovak Academy of Sciences, Košice, Slovakia. She has started the cytogenetic and molecular research of F. magna in Slovakia and motivated the authors to conduct population genetic studies of giant liver fluke.

Margo Pybus is the provincial wildlife disease specialist at the Alberta Fish and Wildlife Division and adjunct professor at the Department of Biological Sciences, University of Alberta, Edmonton, Alberta, Canada. She has managed collection of giant liver fluke samples from all North American enzootic regions, thus enabling us to extend our molecular work on the global scale.

Dušan Rajský is the assistant professor at the Technical University in Zvolen, Slovakia. He was the first researcher to report the presence of fascioloidosis in Slovakia and is managing the sampling of Fascioloides magna *from the Danube floodplain forests natural focus.*

Acknowledgements

The authors would like to acknowledge Marta Špakulová for photographs and drawings used in the current book. This work was supported by the Slovak Grand Agency VEGA (project no. 2/0133/13).

Contents

Introduction

Parasites represent very diverse and in many terms fascinating organisms. The great attention is focused mainly on medically and veterinary important species, which are causative agents of parasitic infections of human and animals. The design of specific and effective diagnostic tools is an important subject of investigation of parasites since their proper diagnostics and identification are the key factors for their effective treatment and control. Besides their clinical importance, parasites are also attractive models for researchers. Host–parasite interactions, immunological responses of hosts to parasitic infections, and mechanisms of adaptation of parasites to host organisms are only a few of many interesting subjects that parasitologists deal with. Invasive parasites, which are very often transmitted into novel territories as a hidden side effect of introduction or translocation of their hosts, are interesting models of investigation, as well.

The model organism of this publication is giant liver fluke, *Fascioloides magna* (Trematoda: Fasciolidae), veterinary important liver fluke of free-living and domestic ruminants. According to the type of final hosts, fascioloidosis can cause different clinical signs and pathological changes, which may occasionally have a lethal effect. It is understandable that giant liver fluke has attracted attention of hunters, farmers, and veterinarians. *Fascioloides magna* is characterized by a wide spectrum of intermediate and final hosts, good ability to adapt to new host species, large spatial distribution, invasive character, and potential to colonize new territories. All these characteristics have been studied for *F. magna* in great detail which led to novel findings related to biology, host–parasite interactions, and distribution of the fluke.

The current publication has a main goal to summarize data on *F. magna* from different aspects of its research and provide a complex survey of results acquired since the first discovery of the parasite in 1875. Chapter 1 is focused on the general information of the parasite, in particular its taxonomic classification, morphology, life cycle, clinical signs, pathology, and therapeutic treatment. Distribution of giant liver fluke and description of North American enzootic regions and European natural foci are provided in Chap. 2. *Fascioloides magna* parasitizes a broad spectrum of final hosts, mainly free-living and domestic ruminants, which can be divided into three categories (definitive, dead-end, and aberrant) according to the host–parasite relationships, pathological changes within the host organism, and ability of fluke to reach maturity, produce eggs, and release them into external environment. Detailed characterizations of final hosts of *F. magna* are provided in Chap. 3. Besides natural infections, this chapter is focused on experimental infections aimed to determine the clinical signs, pathological changes, and immunological responses of final hosts under controlled experimental conditions. Chapter 4 summarizes a spectrum of naturally infected intermediate snail hosts of giant liver fluke in North America and Europe. The results on experimentally infected snails and their potential to serve as the intermediate hosts of giant liver fluke are provided, as well. The latest methods of cellular and molecular biology have been applied in giant liver fluke research only since the 1990s. Chapter 5 summarizes results acquired by modern molecular and immunological approaches. It provides general structure and characterization of ribosomal, mitochondrial genes and microsatellites, and their further utilization in molecular taxonomy and phylogeny of *F. magna*. Data on ultrastructure and karyotype of *F. magna*, results on isoenzyme analyses and studies on excretory/secretory proteins, humoral immune responses, and up-to-date technologies of transcriptome analysis are also included. We hope that this book will provide a useful source of meaningful knowledge and data for anyone professionally engaged in parasitology, including veterinarians, hunters, farmers, and wildlife managers.

Chapter 1
General Information About *Fascioloides magna*

Abstract Giant liver fluke *Fascioloides magna* is a veterinary important liver parasite of free-living and domestic ruminants. This chapter provides general characterization and basic data on the parasite, with focus on its taxonomy, morphology, life cycle, clinical signs, pathology and treatment. Different taxonomic classification and scientific names of the species, and currently accepted taxonomy of *F. magna* are provided in Sect. 1.1. The second part is dealing with morphological description of the parasite, which belongs to the largest flukes worldwide. *Fascioloides magna* utilizes aquatic snails as the intermediate hosts and a wide range of free-living and domestic ruminants as the final hosts. The life cycle of the parasite, divided into four developmental stages, is described in the third subchapter. The fourth part is focused on characterization of clinical signs of fascioloidosis, which are specific for particular type of the final host. Typical pathological changes of *F. magna* infection, described in the fifth subchapter, are fibrous pseudocysts of sedentary adult flukes leading to enlargement of the liver. The last subchapter summarizes the broad spectrum of anthelmintic drugs (e.g. benzimidazoles, salicylanilides, sulphonamides etc.) used for fascioloidosis treatment in different ruminants. Out of them, triclabendazole and rafoxanide proved high efficacy against adult and immature flukes; however, no specific therapeutics are available till now.

Keywords Giant liver fluke · Taxonomy · Morphology · Life cycle · Clinical signs · Pathology · Treatment

1.1 Taxonomic Classification

In spite of the generally accepted North American origin of *Fascioloides magna*, the first case report and description of giant liver fluke originates from Europe from the second half of the 19th century (Swales 1935). In 1875, Italian veterinarian Roberto Bassi described new parasite from the liver of a wapiti stag from the Royal

I. Králová-Hromadová et al., *The Giant Liver Fluke, Fascioloides magna: Past, Present and Future Research*, SpringerBriefs in Animal Sciences,
DOI 10.1007/978-3-319-29508-4_1

Table 1.1 Taxonomic classification of *F. magna* modified according to Jones (2005)

Phylum	Platyhelminthes
Class[a]	Trematoda Rudolphi, 1808
Subclass	Digenea Carus, 1863
Order	Echinostomida La Rue, 1957
Superfamily	Echinostomatoidea Looss, 1899
Family	Fasciolidae Railliet, 1895
Subfamily	Fasciolinae Railliet, 1895
Genus	*Fascioloides* Ward, 1917
Species	*Fascioloides magna* (Bassi, 1875) Ward, 1917
English names	Giant liver fluke, large American liver fluke, deer fluke

[a]class Trematoda belongs to the lineage Neodermata

Park La Mandria in northwestern Italy (Bassi 1875 c.i. Pybus 2001). The species was named as *Distomum magnum* Bassi (1875).

Later on, Charles W. Stiles studied liver flukes from North American cervids and found parasites, which were identical with those described as *Distomum magnum* by Bassi (Stiles and Hassall 1894 c.i. Pybus 2001). Based on these comparisons, Stiles drafted first comprehensive morphological description of the fluke and renamed it as *Fasciola magna* (Swales 1935). In 1895, Stiles noticed similarities between life cycles of *Fasciola magna* and *Fasciola hepatica* (i.e. utilization of aquatic snails as intermediate hosts), and provided description of eggs and miracidium larval stages of the parasite (Stiles and Hassall 1895 c.i. Pybus 2001).

Finally, due to the morphological differences between *Fasciola magna* and other species of the genus *Fasciola* (e.g. the lack of distinct anterior cone and localization of vitellaria in the region ventral to the intestinal branches), Henry B. Ward proposed the new genus *Fascioloides* with the only type species *Fascioloides magna* (Ward 1917) (Table 1.1).

1.2 Morphology

One of the most significant morphological characteristics of *F. magna* is its large body and thick size, due to which it belongs to one of the largest trematodes worldwide. The overall size of adult flukes varies between 40–100 mm of length and 20–35 mm of width (Fig. 1.1); body thickness ranges from 2 to 4.5 mm (Erhardová 1961).

The body is oval or leaf-shaped, dorsoventrally flattened, non-segmented and bilaterally symmetrical (Erhardová-Kotrlá 1971). The body surface is covered by *tegument* with fine spines except for the anterior part of the flukes. The reddish-brown colour of the body is caused by translucent contents of the intestine (Špakulová et al. 2003). Moreover, some internal organs can be visible through the

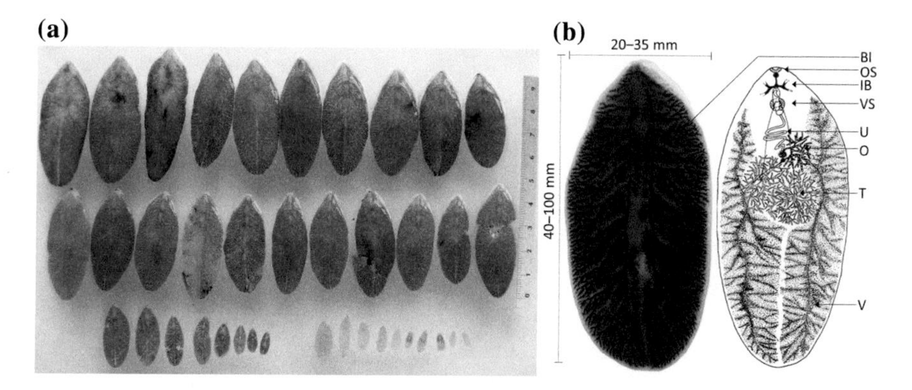

Fig. 1.1 **a** infrapopulation of *F. magna* from red deer from Danube floodplain forests, Slovakia (*Photo* M. Špakulová); **b** morphology of adult *F. magna* (*Photo* E. Bazsalovicsová, *drawing* M. Špakulová); *BI* branched intestine, *OS* oral sucker, *IB* intestine bifurcation, *VS* ventral sucker, *U* uterus, *O* ovarium, *T* testes, *V* vitellaria

body surface, which is underlaid by *musculature* consisting of several layers: the outer circular, intermediate longitudinal and inner diagonal muscles (Erhardová-Kotrlá 1971; Trailović et al. 2015).

The anterior end of the fluke is slightly pointed, while the posterior margin is widely rounded. In the anterior half of the body, two muscular suckers are localized; the *oral sucker* surrounding the mouth opening is usually connected with oral cavity and allows sucking of blood (*hematophagy*). The *ventral sucker* (*acetabulum*) is localized on the ventral side in the first third of the body, 3–4 mm from the oral sucker and serves as attachment organ (Erhardová-Kotrlá 1971).

Internal organs of mature hermaphrodic flukes are present in the parenchyma. The *digestive system* with well-developed oral sucker and the sac-like intestine are differentiated already in the rediae. It is formed by mouth surrounded by the oral sucker, passing to buccal cavity, following by short muscular pharynx and oesophagus. It is bifurcated into two branched intestinal systems, largely extended into many diverticula, which continued along the whole body length (Fig. 1.1) (Erhardová-Kotrlá 1971; Špakulová et al. 2003). The suckers and pharynx contain numerous receptory cells (Erhardová-Kotrlá 1971). Abundantly branched intestine is blindly terminated in the parenchyma and creates a thin epithelium (*gastrodermis*) with the ability of absorption and secretion (Stiles and Hassall 1895 c.i. 2001).

Giant liver fluke has protonephridial *excretory system*, which forms network of excretory canals opened to the outside of body through terminal excretory pore. Basic structures of the excretory system are flame cells deposited in the parenchyma. The *nervous system* of *F. magna* consists of a paired nerve ganglion and nerve cords (longitudinal and transverse) extending throughout the fluke body (Erhardová-Kotrlá 1971).

Adult flukes of *F. magna* are characterized by the presence of male and female reproductive systems in each individual (Fig. 1.1). Sexually mature flukes have one common *genital pore* for both reproductive systems, which is median and immediately pre-acetabular. *Male reproductive organs* consist of two branched *testes* localized closer to the body centre and *vas deferens* opening into the *bursa cirri*. The walls of the *vas deferens* pass into the *vesicula seminalis* and then in the *pars prostatica*. The copulatory organ (ejaculatory duct) terminates in short *cirrus*. Testes are placed side by side almost in the entire second third of the body, but usually one testis may be slightly above the other (Erhardová-Kotrlá 1971).

Female reproductive system consists of lobulated and branched *ovarium*, which is pretesticular, situated slightly to the right side. The *oviduct* surrounded by the Mehlis' gland lies in the middle, and is connected with the *transverse vitelline ducts*. The short Laurer's canal extends from the oviduct. Widely branched *vitellarium* is localized only on the ventral side of the digestive system, and vitelline fields fill lateral regions of body from level of acetabulum to posterior body end. *Oviduct* continues by ball-shaped *uterus* placed in the first third of the body, proceeding to the *bursa cirri* and short *metraterm*.

Uterus is usually filled with a large number of eggs, which possess typical *operculum* on an apical pole (Erhardová-Kotrlá 1971). The peculiar appendage localized opposite the opercular end of the egg is present in variable forms on practically all eggs removed from liver pseudocysts filled with flukes. This appendage is present only on approximately 20 % of eggs normally passed from the final hosts (Swales 1935). The overall length of eggs is 109–175 μm, width is 81–117 μm, depending on the type of the final host and course of infection. The eggs are oval or slightly widened in the centre, yellow or yellowish-brown in colour. They are covered by shell, which is approximately 3.0 μm thick and smooth; however, the germ and vitelline cells are visible through the translucent surface (Swales 1935; Campbell 1961; Erhardová-Kotrlá 1971).

1.3 Life Cycle

The complete life cycle of *F. magna* was described by Swales in 1935. The details of all developmental stages of the life cycle were later specified by Erhardová-Kotrlá (1971). Giant liver fluke has complex life cycle with four stages (Fig. 1.2). The first developmental stage takes place in external environmental conditions, including the phase after dissemination of eggs within the host's faeces into water environment and their development to miracidium. The second stage involves the development of different larval stages (sporocysts, mother and daughter rediae) within the intermediate hosts (aquatic snails). In the third stage, metacercariae develop after release of cercariae from intermediate host in the humid external environment. The fourth stage begins after the ingestion of infective metacercariae by final hosts (e.g. cervids or other ruminants), and continues up to the maturity of adult flukes and production of eggs.

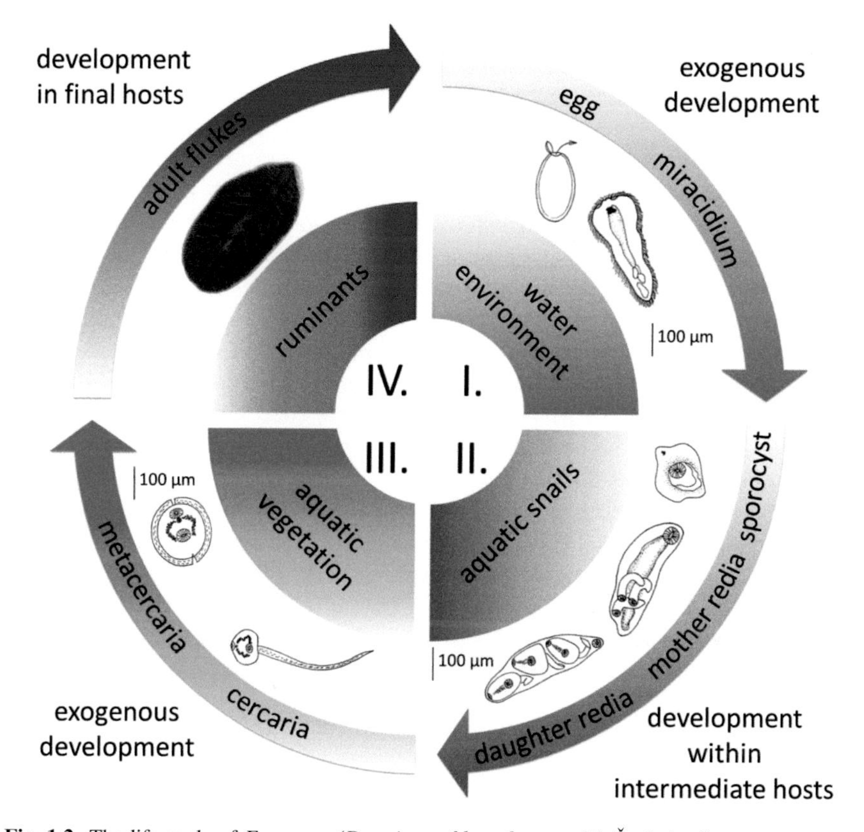

Fig. 1.2 The life cycle of *F. magna* (*Drawings of larval stages* M. Špakulová)

First stage: exogenous development Adult flukes are usually localized in the liver pseudocysts of final hosts, in particular in definitive type of hosts (for details see 1.4 and 1.5). Mature flukes may release up to 4,000 thick-walled operculated *eggs* per day (Swales 1935). The eggs are released with bile into the intestine and leave host's organism along with faeces (Erhardová-Kotrlá 1971). In mature eggs, the process of embryonation results in formation of larval stage, *miracidium*. Complete embryonation takes approximately 35 days (Swales 1935), but this time varies considerably with changes in temperature and moisture (Pybus 2001). In general, the reduction of the temperature prolongs the development (Erhardová-Kotrlá 1971). Low temperatures (<20 °C) retard development, while high temperatures (>34 °C) lead to abnormalities in embryonation and an inability to hatch (Campbell 1961).

In external environmental conditions, a fully developed ciliated larva miracidium hatches from egg by opening the operculum and is released into the water environment during 2–4 weeks (Schwartz et al. 1993). The growing miracidium produces proteolytic enzymes that liberate the egg operculum and allow hatching (Pybus et al. 1991).

Free-living miracidia are fast moving stages, which actively seek intermediate snail hosts and penetrate into the snail's body under the mantle fold on the posterior part of the pulmonary sac (Swales 1935). The penetration into the aquatic snails is facilitated by the secretion of apical gland situated in the anterior part of miracidium. One miracidium has already fully developed six sensory organs, which are connected with the central ganglion (Erhardová-Kotrlá 1971). Miracidia display positive phototaxis and have strong affinity for mucus of lymnaeid snails (Campbell 1961; Erhardová-Kotrlá 1971). If miracidia do not penetrate into the suitable intermediate host, they gradually lose their energy and die. They can survive in humid environment from 10–16 h (Erhardová 1961) up to 1–2 days (Pybus 2001).

Second stage: development within intermediate hosts The development within the intermediate host (or multiplication phase) begins by the creation of new larval stage, *sporocyst*. Miracidia can migrate through the snail's body away from the penetration site. Therefore, sporocysts can sometimes be found in various sites; in the shell cavity or in the shell, in the foot of snail, near the digestive system or in pulmonary cavity. However, they have never been found in the hepatopancreas and kidney (Erhardová-Kotrlá 1971). Transformation of miracidia into the sporocyst takes about 8–10 h after the penetration (Swales 1935; Erhardová 1961; Schwartz et al. 1993). In the stage of sporocyst, only muscular pharynx, buccal cavity and rudimentary oesophagus are developed; no other internal organs are evolved. Sporocysts are able to form two types of rediae, which enter the snail's tissue (Schwartz et al. 1993).

At first, sporocysts change their form and develop into *mother rediae*. Each sporocyst usually contains only one mother redia and 4–6 germ cells. Mother rediae elongate their body, actively move and escape from sporocysts by rupturing their wall. They have completely developed digestive organs; mouth, very large muscular pharynx, oesophagus and gut. Mother rediae migrate through the tissue of the snail; they can be found mainly in the kidney, female reproductive organs, pulmonary cavity and near the anal pore. Each mother redia contains 4–6 light-yellow coloured *daughter rediae* (Erhardová 1961), which gradually develop in growing mother rediae. They have very similar digestive system comparing to mother rediae, involving mouth, pharynx, oesophagus and gut. Smaller muscular pharynx of daughter rediae is differential feature from mother rediae. The shape of body of daughter redia is divided into two parts; the anterior is larger and wider, while the posterior one is shorter and narrower. Within each daughter redia, the next larval stage (*cercaria*) develops in various numbers; the highest number is usually six cercariae in one daughter redia (Erhardová-Kotrlá 1971). An apparent difference between mother and daughter rediae is the retention of the strong collar in the anterior part of the redia. In daughter rediae containing cercariae, no indication of collar is observed, while in mother rediae, it is formed by a simple fold in the wall, probably as a result of growing and stretching cercariae (Swales 1935).

Third stage: exogenous development The cercariae emerge from the daughter rediae and they usually migrate into the hepatopancreas and reproductive organs of snails, where they complete their development. Mature cercariae represent free-living larval stage, which persist in external environment after leaving snail.

They are very active inside the daughter rediae, where they have been formed. The light-yellow coloured cercariae are very similar by their body construction with adult stages of fluke; anterior portion is heart-shaped and wide, while the posterior one represents one long tail. Their digestive system is formed by mouth, muscular pharynx, short oesophagus, intestine bifurcated into two branches and caecum. Some kind of excretory system and rudimentary basis of reproductive organs are also developed (Swales 1935; Erhardová-Kotrlá 1971).

After the complex of multiplication processes in the intermediate host, several larval generations are produced during approximately 2.5 months. As a result, about 1,000 cercariae are released from infected snail (Swales 1935; Erhardová 1961). Development in snails depends mainly on physical conditions (e.g. temperature, moisture etc.), and type or species of intermediate hosts (Pybus 2001). Finally, the cercariae move from the snail's tissue back into the water, migrate a short distance, encyst on the surface of an aquatic vegetation and develop into the *metacercariae* (Schwartz et al. 1993). Metacercariae represent the stage infectious for final hosts; they remain infectious during 2–2.5 months fixed on submergent or emergent vegetation, particularly in cold water. The dark-brown metacercariae are covered by the wall, which is formed by two layers; thinner inner and thicker outer (Erhardová-Kotrlá 1971; Schwartz et al. 1993). Metacercariae-infected herbage may be ingested by domestic or free-living ruminants (Foreyt and Parish 1990), mainly in two primary transmission periods, in the late summer and fall, and in the spring (Erhardová-Kotrlá 1971).

Fourth stage: development in the final hosts After ingestion of metacercariae, activated larva penetrates the intestinal wall of its final host, migrate along the ventral aspect of the peritoneal cavity, and then penetrates the liver through the Glisson's capsule, where they slowly grow and develop into adults (Pybus 2001). In final hosts, flukes mature approximately 30 weeks after infection (Foreyt and Todd 1976a). Localization of flukes depends on the type of the final host (definitive, aberrant, dead-end; for details see Chap. 3). In definitive host, *F. magna* occurs in thin-walled fibrous pseudocysts within the liver parenchyma usually in pairs, but occasionally also in higher numbers (Foreyt et al. 1977; Schwartz et al. 1993; Pybus 2001). Hermaphroditic helminths in general prefer cross-fertilization; however, self-fertilization may occur in absence of available partner (Špakulová et al. 1996).

In dead-end hosts, thick-walled encapsulation of flukes was observed, while for aberrant hosts, excessive wandering of immature flukes and lack of encapsulation are typical. Single immature fluke may migrate through the hepatic parenchyma up to one year before becoming encapsulated with other fluke (Foreyt et al. 1977; Mulvey et al. 1991). Such immature flukes may migrate aimlessly and destructively through the organs of abdominal or thoracic cavities. Prepatent period of *F. magna* in ruminants ranges from three (Erhardová-Kotrlá 1971) to seven months (Swales 1935; Foreyt and Todd 1976a). Adult flukes survive in liver of final hosts at least five years (Erhardová-Kotrlá 1971).

1.4 Clinical Signs

Clinical signs caused by *F. magna* infection strongly depend on the type of final host (definitive, aberrant, dead-end; for details see Chap. 3). Infection in the definitive hosts (e.g. white-tailed deer, wapiti, red deer etc.) is usually well tolerated (except for young animals, or animals of lower fitness) and *F. magna* is not considered as a serious pathogen in these cervids (Swales 1935; Griffiths 1962; Foreyt and Todd 1976a). However, some clinical signs, such as lethargy, poor appetite, anorexia, anemia, depression and weight loss, may occur (Foreyt 1992, 1996a). In occasional cases, fascioloidosis can lead to death of definitive host, as was reported e.g. in naturally infected white-tailed deer (Pursglove et al. 1977), red deer (Balbo et al. 1987), and also in experimentally infected wapiti (Foreyt 1996b) or mule deer (Foreyt 1992).

Contrary to definitive hosts, infections in aberrant (e.g. sheep, goat, roe deer) and dead-end hosts (e.g. cattle, moose, sika deer) display different clinical signs and may more often cause a lethal effect. The course of infection in these types of hosts could also be subclinical and coprological examinations may not provide reliable results (Foreyt and Todd 1976b; Stromberg et al. 1983). For instance, goats, sheep and llama apparently did not exhibit initial clinical signs of infection, but in some cases, infected animals show signs of lethargy and weakness shortly before death (Foreyt 1990, 1996a). Mortality usually occurs within 4–6 months post-infection (Swales 1935; Erhardová-Kotrlá and Blažek 1970; Erhardová-Kotrlá 1971; Foreyt and Leathers 1980), and may be associated with acute peritonitis before or after migrating larvae reach the liver. Since fascioloidosis in domestic ruminants may cause significant economic losses, monitoring of farmed animals and compliance of preventive measures are highly important in order to prevent infections in cattle, sheep and goats (Lanfranchi et al. 1985).

1.5 Pathology

Pathological changes caused by *F. magna* infection also depend on the type of its ruminant hosts, with different tolerance to fascioloidosis (Pybus 2001). In definitive hosts, fibrous encapsulated pseudocysts of sedentary adult flukes (Fig. 1.3), which lead to pathological enlargement of the liver, are typical. The liver has usually rounded margins and fibrous tags on the serosa (Pybus 2001). An enlarged grey liver with irregular grooves, fibrin and scattered diffuse foci of black pigment on the surface, a lot of different cystic spaces filled with brownish mucous fluid, and changes in the liver tissue typical for cirrhosis, were also observed (Karamon et al. 2015). Perivascular inflammation is generally not detectable (Foreyt and Todd 1979).

The infected liver is predominantly characterized by primary lesions associated with mechanical damage due to migrating immature flukes. After flukes'

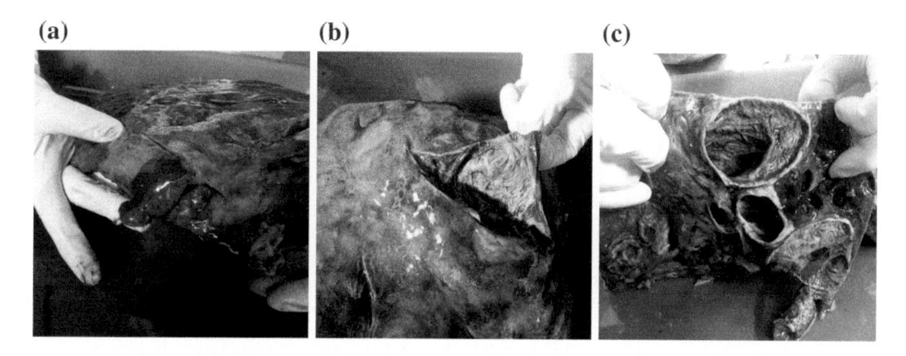

Fig. 1.3 **a** adult *F. magna* in liver of infected red deer from Danube floodplain forests, Slovakia; **b** pathological changes of infected liver; **c** details on fibrous capsules in the liver parenchyma (*Photos* M. Špakulová and E. Bazsalovicsová)

consumption of blood and its components, particularly erythrocytes, streaks of black pigment may be seen. They occur in abdominal or thoracic organs, especially in the liver. Black pigment accumulates in hepatic cells either on the serosal surface or throughout the liver parenchyma (Swales 1935; Pybus 2001). Black spots of different size may be visible on the omentum, peritoneum, pleura and cranial part of lungs (Karamon et al. 2015). Black or dark-green pigment belongs to the group of hematins and is produced in the intestine of immature and adult flukes as a by-product of feeding on blood (Campbell 1960; Blažek and Gilka 1970). The presence of hematin is typical only for giant liver fluke; there is no evidence of its occurrence in parasitic infections caused by other flukes (Chroust 1987).

Thin-walled pseudocysts usually arise as a result of an immune defense mechanism against the migration of young flukes throughout the hepatic parenchyma and are filled with a dark-green liquid. With the gradual development of capsule around the flukes, the surrounding liver parenchyma is destroyed due to pressure atrophy (Swales 1935, 1936). Capsules are of host origin and are an apparent attempt to prevent further migration of flukes within the liver parenchyma (Pybus 2001).

Cysts with adult flukes are situated in the liver parenchyma and are opened to the biliary system (Conboy and Stromberg 1991). Giant liver flukes usually occur in pairs (Erhardová-Kotrlá 1971), although in some cases more than two flukes can be present in one pseudocyst (Špakulová et al. 2003). The size of pseudocysts is variable (average size 50–100 mm); it depends on the amount of accumulated fluid and detritus, but also on the number and size of flukes enclosed in cyst (Swales 1935, 1936).

In aberrant hosts, pathological changes are mainly characterized by excessive wandering of immature flukes and lack of encapsulation. In addition to necrosis throughout the liver, perforation of the hepatic capsule or penetration into various abdominal and pleural organs (most frequently lungs) may also occur (Foreyt and Todd 1976b; Foreyt and Leathers 1980). In dead-end hosts, fibrosis and thick-walled encapsulation of flukes was observed. In some cases, even chronic calcification of pseudocysts may occur (Pybus 2001). The presence of black pigmentation of various tissues is one of the initial macroscopical diagnostic markers of *F. magna* infections in this type of hosts (Špakulová et al. 2003). Higher number of flukes and prolonged infections cause more extensive histopathological changes in liver parenchyma of infected ruminants (Pybus 2001).

1.6 Therapeutic Treatment

Therapeutic treatment of fascioloidosis in domestic ruminants is feasible as a part of on-going individual herd management programs (Pybus 2001). An important role plays the way of housing animals in farms, pasture rotation and the frequency of animals in pastures, improvement of animal zoo-hygienic conditions and general animal welfare. Therefore, the treatment of domestic ruminants appears to be more effective due to possibility to control the dosage and administration of anthelmintic drugs. It is fully copying the knowledge on pharmacological anthelmintic treatment of liver fluke *Fasciola hepatica* (Trematoda; Fasciolidae), species which is closely related to *F. magna* (Mas-Coma 2005). Due to this fact, pharmacological treatment, as known for *F. hepatica* (Fairweather and Boray 1999), was adopted in treatment of *F. magna*. However, *F. magna* infections are in most cases difficult to treat because flukes are not localized directly in the bile ducts as in *F. hepatica* infections. As a result, most anthelmintic drugs effective against *F. hepatica* do not work well against *F. magna* (Foreyt and Todd 1976b).

Treatment of *F. magna* in free-living ruminants differs from that of domestic ones. Anthelmintic drugs are administrated to animals through the feeding mixtures in winter seasons. In such cases, an exact dose of anthelmintics can not be controlled and treatment of cervids is mostly unfeasible (Pybus 2001). The main reason is that mixing of drugs with salt might limit the amount of mixture that can be eaten in one visit to the feeding table by the dominant deer, and leaving enough to treat the inferior ones (Janicki et al. 2005). Direct treatment of cervids is possible only in situations of their translocations from enzootic areas to husbandry (Pybus 2001). Natural populations of cervids are largely difficult to treat due to inability of drugs to penetrate into liver pseudocysts (Rajský et al. 2002). The side effect of treatment is presence of drugs in muscles or other tissues of cervids, what is often less desirable than drug-free animal with good tolerance of fascioloidosis. Table 1.2 summarizes data on anthelmintic drugs used in fascioloidosis treatment of free-living and domestic ruminants in North America and Europe.

In the long-term history of fascioloidosis treatment, several groups of anthelmintics with different type of agents were administrated in veterinary and husbandry practices (see Table 1.2 and references therein). Some anthelmintic groups were proved to be effective against mature and immature stages of *F. magna* in different type of final hosts (Foreyt and Todd 1974; Balbo et al. 1987; Qureshi et al. 1989). High level of efficacy (63–100 %) was detected for *triclabendazole* (*benzimidazoles*) with best results against both forms of flukes (adult and immature) in

Table 1.2 Summary of anthelmintic drugs used in fascioloidosis treatment of free-living and domestic ruminants

Group of anthelmintics	Agent	Molecular formula	Dose mg/kg	Efficacy %	Against A/I	Final host	References
Benzimidazoles	Triclabendazole	$C_{14}H_9Cl_3N_2OS$	10	100	A, I	White-tailed deer	Qureshi et al. (1989)
			11	63	A, I	White-tailed deer	Qureshi et al. (1994)
			50–60	98	A	Rocky Mountain elk	Pybus et al. (1991)
			50–60	90	I	Rocky Mountain elk	Pybus et al. (1991)
			6–12	77–88	A, I	Cattle	Craig and Huey (1984)
			20	99	I	Goat	Foreyt (1989)
	Albendazole	$C_{12}H_{15}N_3O_2S$	11–54	38	A, I	White-tailed deer	Foreyt and Drawe (1978)
			5–17	82–84	A	White-tailed deer	Qureshi et al. (1990)
			17–46	67	A	White-tailed deer	Foreyt and Drawe (1985)
			17–46	89	I	White-tailed deer	Foreyt and Drawe (1985)
			5–15	70	n.i.	Sheep	Stromberg et al. (1983)
			15–45	94–99	n.i.	Cattle	Ronald et al. (1979)
Salicylanilides	Rafoxanide	$C_{19}H_{11}Cl_2I_2NO_3$	12–25	75	I	White-tailed deer	Foreyt and Todd (1976b)
			10	n.i.	A	Red deer	Balbo et al. (1987)
			15	98	n.i.	Roe deer	Chroust (1987)
			10–15	100	A, I	Cattle	Foreyt and Todd (1974)
	Oxyclozanide	$C_{13}H_6C_{15}NO_3$	13–29	100	A	White-tailed deer	Foreyt and Todd (1973)
			7–15	27	A, I	Cattle	Foreyt and Todd (1974)
	Closantel	$C_{22}H_{14}Cl_2I_2N_2O_2$	7.5–15	94–98	A, I	Sheep	Stromberg et al. (1985)
Sulphonamides	Clorsulon	$C_8H_8Cl_3N_3O_4S_2$	12–30	92	A	White-tailed deer	Foreyt and Drawe (1985)
			12–30	80	I	White-tailed deer	Foreyt and Drawe (1985)
			21	75–100	I	Cattle	Foreyt (1988)
			21	92	I	Sheep	Foreyt (1988)

(continued)

Table 1.2 (continued)

Group of anthelmintics	Agent	Molecular formula	Dose mg/kg	Efficacy %	Against A/I	Final host	References
Halogenated phenols	Hexachlorophene	$C_{13}H_6Cl_6O_2$	12–26	50	A	White-tailed deer	Foreyt and Todd (1976b)
			12–26	0	I	White-tailed deer	Foreyt and Todd (1976b)
	Nitroxynil	$C_7H_3IN_2O_3$	11–24	0	A	White-tailed deer	Foreyt and Todd (1976b)
			11–24	50	I	White-tailed deer	Foreyt and Todd (1976b)
	Bithionolsulfoxide	$C_{12}H_6Cl_4O_3S$	40–50	100	A	Cattle	Chroustová et al. (1980)
Phenoxyalkanes	Diamphenetide	$C_{20}H_{24}N_2O_5$	255–280	0	A, I	White-tailed deer	Foreyt and Todd (1976b)
			140	n.i.	A, I	Red deer	Balbo et al. (1987)

A adult *F. magna*, *I* immature *F. magna*, *n.i.* not indicated in the respective literature white-tailed deer, Rocky Mountain elk and cattle (see Table 1.2 and references therein). Consequently, several authors recommended triclabendazole as the best choice for fascioloidosis treatment (Craig and Huey 1984; Foreyt 1989; Pybus et al. 1991; Qureshi et al. 1989, 1994). Slightly lower efficacy (38–99 %) was observed for *albendazole* (*benzimidazoles*), which was applied for fascioloidosis treatment in white-tailed deer, sheep and cattle (see Table 1.2 and references therein).

According to the other studies on fascioloidosis therapy *salicylanilides* and *sulphonamides* were also highly effective in free-living and domestic ruminants (Table 1.2). *Rafoxanide* (salicylanilides) has high efficacy (100 %) against both forms in cattle (Foreyt and Todd 1974) and was also effective against giant liver flukes in free-living ruminants, e.g. white-tailed deer (75 %; Foreyt and Todd 1976b) and roe deer (98 %; Chroust 1987). *Clorsulon* (sulphonamides) is active mainly against immature flukes parasitizing cattle and sheep (Foreyt 1988), with rather high efficacy (80–92 %) against adult and immature flukes infecting white-tailed deer (Foreyt and Drawe 1985).

Out of *halogenated phenols*, *hexachlorophene* and *bithionolsulfoxide* were proved to be effective against adult flukes in white-tailed deer and cattle, respectively (Foreyt and Todd 1976b; Chroustová et al. 1980). *Nitroxynil* was efficient only against immature flukes in white-tailed deer (Foreyt and Todd 1976b). *Diamphenethide* (*phenoxyalkanes*) fed in medicated pellets effectively controlled *F. magna* infection in captive red deer (Balbo et al. 1987); however, authors did not declare exact efficacy of the drug. In contrast, diamphenethide used in higher dose was not effective either in adults or in immature flukes in white-tailed deer (Foreyt and Todd 1976b).

Until recently, nothing has been known about the metabolism of anthelmintics in *F. magna*. The latest study of Prchal et al. (2015) was focused on determination of the activities of drug-metabolism enzymes in *F. magna* and the metabolism of selected benzimidazoles (triclabendazole, albendazole, mebendazole) and salicylanilides (rafoxanide, closantel), which are commonly used to control fascioloidosis. Specific activities of several drug-metabolizing enzymes (e.g. peroxidase, catalase, glutathione peroxidase, flavine monooxygenase, UDP-glucosyl transferase etc.) were found in subcellular fractions.

The results showed that giant liver fluke is able to oxidize albendazole and reduce mebendazole in vitro; however, it can not oxidize triclabendazole. Ex vivo cultivation of living adult flukes with anthelmintics confirmed the ability of parasites to oxidize albendazole to albendazole sulphoxide and to reduce mebendazole. Concerning the salicylanilides, no metabolites of rafoxanide and closantel formed by *F. magna* were detected. It was concluded, that *F. magna* possess the active xenobiotic-metabolizing system, but it is not able to mediate sufficient protection against anthelmintic drugs (Prchal et al. 2015).

Comparing to other veterinary important parasitoses, fascioloidosis is responsible for generally lower economic consequences. Probably due to this fact, no specific pharmacological therapeutics are available till now. Therefore, preventive measures are of high importance. Particularly high risk represents the feeding with hay from meadows, where are commonly found either infected free-living ruminants or aquatic snails, intermediate hosts of *F. magna*. The suitable alternatives seem to be timely reduction of parasite spreading by physical methods (e.g. drainage or drying of pastures), application of molluscicides to grassland or introduction of competitive species of snails, which eliminate intermediate hosts in habitat (Novobilský and Koudela 2005).

References

Balbo T, Lanfranchi P, Rossi L, Meneguz PG (1987) Health management of a red deer population infected by *Fascioloides magna* (Bassi, 1875) Ward, 1917. Ann Fac Med Vet Torino 32:1–13

Bassi R (1875) Sulla cachessia ittero-verminosa, o marciaia, causta dei Cervi, causata dal *Distomum magnum*. Il Medico Veterinario 4:497–515. Cited in Pybus MJ (2001) Liver flukes. In: Samuel WM, Pybus MJ, Kocan AA (eds) Parasitic diseases of wild mammals, 2nd edn. Iowa State University Press, Ames

Blažek K, Gilka F (1970) Notes of the nature of the pigment of the trematode *Fascioloides magna*. Folia Parasitol 17:165–170

Campbell WC (1960) Nature and possible significance of the pigment in fascioloidiasis. J Parasitol 46:769–775

Campbell WC (1961) Notes of the egg and miracidium of *Fascioloides magna* (Trematoda). T Am Microsc Soc 80:308–319. doi:10.2307/3223642

Chroust K (1987) Současný stav a možnosti tlumení motolice obrovské (*Fascioloides magna*) u zvěře. Veterinářství 37:514–515 (in Czech)

Chroustová E, Hůlka J, Jaroš J (1980) Prevence a terapie fascioloidózy skotu bithionolsulfoxidem. Vet Med (Praha) 25:557–563 (in Czech)

Conboy GA, Stromberg BE (1991) Hematology and clinical pathology of experimental *Fascioloides magna* infection in cattle and guinea pigs. Vet Parasitol 40:241–255

Craig TM, Huey RL (1984) Efficacy of triclabendazole against *Fasciola hepatica* and *Fascioloides magna* in naturally infected calves. Am J Vet Res 45:1644–1645

Erhardová B (1961) Vývojový cyklus motolice obrovské *Fasciola magna* v podmínkách ČSSR. Zool listy 10:9–16 (in Czech)

Erhardová-Kotrlá B (1971) The occurrence of *Fascioloides magna* (Bassi, 1875) in Czechoslovakia. Czechoslovak Academy of Sciences, Prague

Erhardová-Kotrlá B, Blažek K (1970) Artificial infestation caused by the fluke *Fascioloides magna*. Acta Vet Brno 39:287–295

Fairweather I, Boray JC (1999) Fasciolicides: efficacy, actions, resistance and its management. Vet J 158:81–112. doi:10.1053/tvjl.1999.0377

Foreyt WJ (1988) Evaluation of clorsulon against immature *Fascioloides magna* in cattle and sheep. Am J Vet Res 49:1004–1006

Foreyt WJ (1989) Efficacy of triclabendazole against experimentally induced *Fascioloides magna* infections in sheep. Am J Vet Res 50:431–432

Foreyt WJ (1990) Domestic sheep as a rare definitive host of the large American liver fluke *Fascioloides magna*. J Parasitol 76:736–739

Foreyt WJ (1992) Experimental *Fascioloides magna* infections of mule deer (*Odocoileus hemionus hemionus*). J Wildl Dis 28:183–187. doi:10.7589/0090-3558-28.2.183

Foreyt WJ (1996a) Susceptibility of bighorn sheep (*Ovis canadensis*) to experimentally-induced *Fascioloides magna* infections. J Wildl Dis 32:556–559. doi:10.7589/0090-3558-32.3.556

Foreyt WJ (1996b) Mule deer (*Odocoileus hemionus*) and elk (*Cervus elaphus*) as experimental definitive hosts for *Fascioloides magna*. J Wildl Dis 32:603–606. doi:10.7589/0090-3558-32.4.603

Foreyt WJ, Drawe DL (1978) Anthelmintic activity of albendazole in white-tailed deer. Am J Vet Res 39:1901–1903

Foreyt WJ, Drawe DL (1985) Efficacy of clorsulon and albendazole against *Fascioloides magna* in naturally infected white-tailed deer. J Am Vet Med Assoc 187:1187–1188

Foreyt WJ, Leathers CW (1980) Experimental infection of domestic goats with *Fascioloides magna*. Am J Vet Res 41:883–884

Foreyt WJ, Parish S (1990) Experimental infection of liver flukes (*Fascioloides magna*) in a llama (*Lama glama*). J Zoo Wildl Med 21:468–470

Foreyt WJ, Todd AC (1973) Action of oxyclozanide against adult *Fascioloides magna* (Bassi 1875) infections in white-tailed deer. J Parasitol 59:208–209

Foreyt WJ, Todd AC (1974) Efficacy of rafoxanide and oxyclozanide against *Fascioloides magna* in naturally infected cattle. Am J Vet Res 35:375–377

Foreyt WJ, Todd AC (1976a) The development of the large American liver fluke, *Fascioloides magna*, in white-tailed deer, cattle, and sheep. J Parasitol 62:26–32

Foreyt WJ, Todd AC (1976b) Liver flukes in cattle: prevalence, distribution and experimental treatment. Vet Med Small Anim Clin 71:816–822

Foreyt WJ, Todd AC (1979) Selected clinicopathologic changes associated with experimentally induced *Fascioloides magna* infection in white-tailed deer. J Wildl Dis 15:83–89

Foreyt WJ, Samuel WM, Todd AC (1977) *Fascioloides magna* in white-tailed deer (*Odocoileus virginianus*): observation of the pairing tendency. J Parasitol 63:1050–1052. doi:10.2307/3279843

Griffiths HJ (1962) Fascioloidiasis of cattle, sheep and deer in Northern Minnesota. J Am Vet Med Assoc 140:342–347

Janicki Z, Konjević D, Severin K (2005) Monitoring and treatment of *Fascioloides magna* in semi-farm red deer husbandry in Croatia. Vet Res Commun 29:83–88. doi:10.1007/s11259-005-0027-z

Jones A (2005) Family Fasciolidae Railliet, 1895. In: Jones A, Braz RA, Gibson DI (eds) Keys to the Trematoda. CABI Publishing, Wallingford

Karamon J, Larska M, Jasik A, Sell B (2015) First report of the giant liver fluke (*Fascioloides magna*) infection in farmed fallow deer (*Dama dama*) in Poland—pathomorphological changes and molecular identification. Bull Vet Inst Pulawy 59:339–344. doi:10.1515/bvip-2015-0050

Lanfranchi P, Tolari F, Forletta R, Meneguz PG, Rossi L (1985) The red deer as reservoir of parasitic and infectious pathogens for cattle. Ann Fac Med Vet Torino 30:1–17

Mas-Coma S (2005) Epidemiology of fascioliasis in human endemic areas. J Helminthol 79:207–216. doi:10.1079/JOH2005296

Mulvey M, Aho JM, Lydeard C, Leberg PL, Smith MH (1991) Comparative population genetic structure of a parasite (*Fascioloides magna*) and its definitive host. Evolution 45:1628–1640. doi:10.2307/2409784

Novobilský A, Koudela B (2005) Treatment and control of *Fascioloides magna* infection in cervids—review. Veterinářství 55:98–102

Prchal L, Vokřál I, Kašný M, Rejšková L, Zajíčková M, Lamka J, Skálová L, Lecová L, Szotáková B (2015) Metabolism of drugs and other xenobiotics in giant liver fluke (*Fascioloides magna*). Xenobiotica. doi:10.3109/00498254.2015.1060370

Pursglove SR, Prestwood AK, Ridgeway TR, Hayes FA (1977) *Fascioloides magna* infection in white-tailed deer of southeastern United States. J Am Vet Med Assoc 171:936–938

Pybus MJ (2001) Liver flukes. In: Samuel WM, Pybus MJ, Kocan AA (eds) Parasitic diseases of wild mammals, 2nd edn. Iowa State University Press, Ames

Pybus MJ, Onderka DK, Cool N (1991) Efficacy of triclabendazole against natural infections of *Fascioloides magna* in wapiti. J Wild Dis 27:599–605. doi:10.7589/0090-3558-27.4.599

Qureshi T, Craig TM, Drawe DL, Davis DS (1989) Efficacy of triclabendazole against fascioloidiasis (*Fascioloides magna*) in naturally infected white-tailed deer (*Odocoileus virginianus*). J Wild Dis 25:378–383. doi:10.7589/0090-3558-25.3.378

Qureshi T, Davis DS, Drawe DL (1990) Use of albendazole in feed to control *Fascioloides magna* infections in captive white-tailed deer (*Odocoileus virginianus*). J Wildl Dis 26:231–235. doi:10.7589/0090-3558-26.2.231

Qureshi T, Drawe DL, Davis DS, Craig TM (1994) Use of bait containing triclabendazole to treat *Fascioloides magna* infections in free ranging white-tailed deer. J Wildl Dis 30:346–350. doi:10.7589/0090-3558-30.3.346

Rajský D, Čorba J, Várady M, Špakulová M, Cabadaj R (2002) Control of fascioloidosis (*Fascioloides magna* Bassi, 1875) in red deer and roe deer. Helminthologia 39:67–70

Ronald NC, Craig TM, Bell RR (1979) A controlled evaluation of albendazole against natural infections with *Fasciola hepatica* and *Fascioloides magna* in cattle. Am J Vet Med 40:1299–1300

Schwartz WL, Lawhorn DB, Montgomery E (1993) *Fascioloides magna* in a feral pig. Swine Health Prod 1:27

Stiles CW, Hassall A (1894) The anatomy of the large American fluke (*Fasciola magna*), and a comparison with other species of the genus *Fasciola*. J Comp Med Vet Arch 15:161–178, 225–243, 299–313, 407–417, 457–462. Cited in Pybus MJ (2001) Liver flukes. In: Samuel WM, Pybus MJ, Kocan AA (eds) Parasitic diseases of wild mammals, 2nd edn. Iowa State University Press, Ames

Stiles CW, Hassall A (1895) The anatomy of the large American liver fluke (*Fasciola magna*) and a comparison with other species of the genus *Fasciola*. J Comp Med Vet Arch 16:139–147, 213–222, 277–282. Cited in Pybus MJ (2001) Liver flukes. In: Samuel WM, Pybus MJ, Kocan AA (eds) Parasitic diseases of wild mammals, 2nd edn. Iowa State University Press, Ames

Stromberg BE, Schlotthauer JC, Karns PD (1983) Current status of *Fascioloides magna* in Minnesota. Minnesota Veterinarian 23:8–13

Stromberg BE, Conboy GA, Hayden DW, Schlotthauer JC (1985) Pathophysiologic effects of experimentally induced *Fascioloides magna* infection in sheep. Am J Vet Res 46:1637–1641

Swales WE (1935) The life cycle of *Fascioloides magna* (Bassi, 1875), the large liver fluke of ruminants, in Canada. Can J Res 12:177–215. doi:10.1139/cjr35-015

Swales WE (1936) Further studies on Fascioloides magna (Bassi 1875 Ward, 1917, as a parasite of ruminants. Can J Res 14:83–95

Šnábel V, Hanzelová V, Mattiucci S, D'Amelio S, Paggi L (1996) Genetic polymorphism in *Proteocephalus exiguus* shown by enzyme electrophoresis. J Helminthol 70:345–349. doi:10.1017/S0022149X00015649

Špakulová M, Rajský D, Sokol J, Vodňanský M (2003) Cicavica obrovská (*Fascioloides magna*). Významný pečeňový parazit prežúvavcov. PaRPRESS, Bratislava (in Slovak)

Trailović SM, Marinković D, Trailović JN, Milovanović M, Marjanović DS, Aničić IR (2015) Pharmacological and morphological characteristics of the muscular system of the giant liver fluke (*Fascioloides magna*—Bassi 1875). Exp Parasitol 159:136–142. doi:10.1016/j.exppara.2015.09.012

Ward HB (1917) On the structure and classification of North American parasitic worms. J Parasitol 4:1–12

Chapter 2
Distribution of *Fascioloides magna*

Abstract Giant liver fluke has established permanent natural foci on two continents. North America represents the original continent of the parasite occurrence, while Europe is the continent where *F. magna* was introduced along with its cervid hosts. In North America, *F. magna* occurs in five enzootic regions across the United States and southern Canada: (1) the northern Pacific coast; (2) the Rocky Mountain trench; (3) the Great Lakes region; (4) northern Quebec and Labrador; and (5) Gulf coast, lower Mississippi, and southern Atlantic seaboard. In Europe, giant liver fluke has established three permanent natural foci: (1) La Mandria Regional Park in the northern Italy; (2) Czech Republic and southwestern Poland; and (3) Danube floodplain forests, involving Austria, Slovakia, Hungary, Croatia and Serbia. This chapter summarizes details on *F. magna* distribution in all North American enzootic regions and European natural foci. Besides permanent foci, sporadic findings of the parasite have been reported throughout the world. Occasional findings very probably represented only single detection of the parasite introduced to the particular region without further establishment of the permanent focus.

Keywords Giant liver fluke · Distribution · Biological invasion · North America · Europe · Enzootic region · Natural focus

2.1 North America

North America has been recognized as the original continent of giant liver fluke. To date, *F. magna* occurs in five enzootic regions across the United States and southern Canada: (1) the northern Pacific coast (NPC); (2) the Rocky Mountain trench (RMT); (3) the Great Lakes region (GLR); (4) northern Quebec and Labrador (NQL); and (5) Gulf coast, lower Mississippi, and southern Atlantic seaboard (SAS) (Fig. 2.1; Pybus 2001). The US states and Canadian provinces with confirmed natural infections of free-living and domestic ruminants are illustrated on Fig. 2.2. Details on geographic localities, final hosts and prevalence of fascioloidosis in all North American enzootic regions are provided in Table 2.1.

I. Králová-Hromadová et al., *The Giant Liver Fluke, Fascioloides magna: Past, Present and Future Research*, SpringerBriefs in Animal Sciences,
DOI 10.1007/978-3-319-29508-4_2

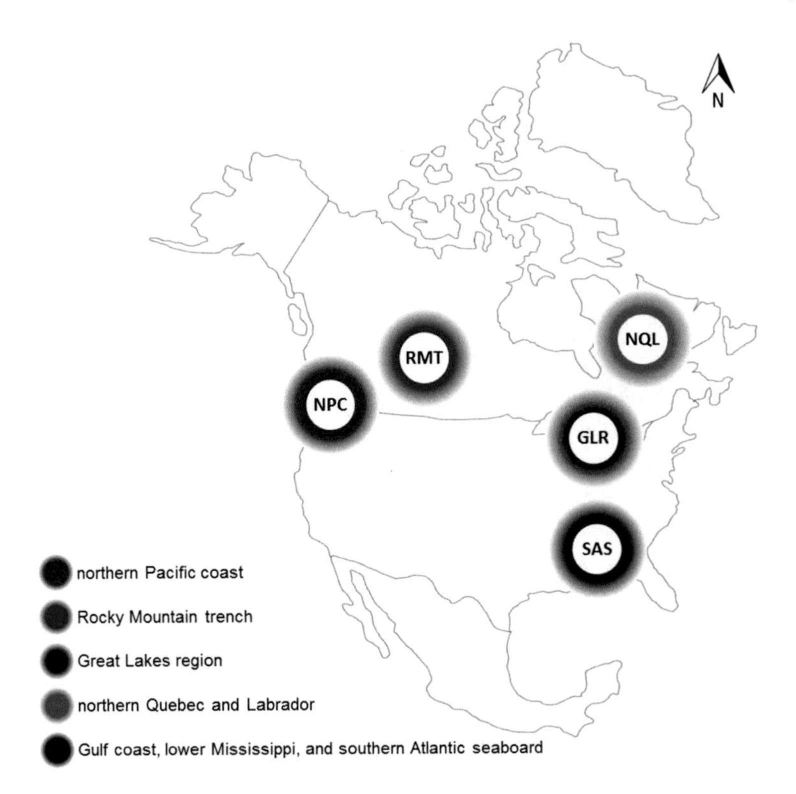

Fig. 2.1 Schematic presentation of North American enzootic regions of *F. magna*

Distribution of *F. magna* in North America has been influenced by a natural migration and human-directed translocations of cervids. Definitive hosts, such as white-tailed deer, wapiti and caribou, have played major role in maintaining fascioloidosis in the natural environment and its further spread into currently recognized enzootic regions. It is generally accepted that *F. magna* have co-evolved with ancestral *Odocoileus* sp. and was originally widespread in white-tailed deer in major wetland habitats throughout North America (Pybus 2001). Interrelationships between white-tailed deer and *F. magna* are finely tuned; the number of flukes within an individual deer is usually limited, allowing maintenance of the parasite population, but not dispersing it beyond its foci. Wapiti and caribou, sympatric with white-tailed deer, encountered *F. magna* in overlapping contaminated wetland habitats. In contrast to the situation in white-tailed deer, potential translocation of liver flukes in wapiti is higher due to increased *F. magna* eggs production and subsequent release into the environment (Pybus 2001).

The population of white-tailed deer declined steadily in 16th and 20th centuries; this process proceeded from the east towards the west of North America and resulted in the disappearance of the giant liver fluke over much of its former range (Pybus 2001 and references therein). However, as claimed by Pybus (2001),

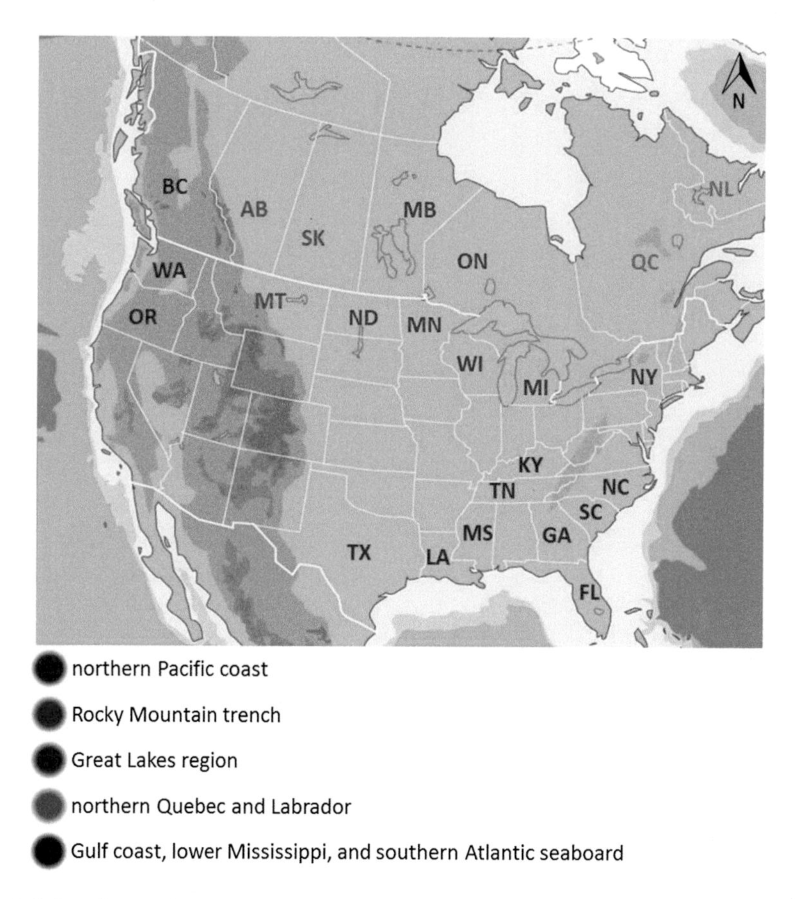

Fig. 2.2 Details on US states and Canadian provinces with confirmed natural infections of *F. magna* in free-living and domestic ruminants (codes are explained in Table 2.1; map downloaded from www.johomaps.com)

populations of the fluke may have remained in three primary refuge areas, particularly in: (i) RMT region in persistent wapiti populations; (ii) SAS region with continued occurrence of white-tailed deer; and (iii) NQL region, where caribou remained unavailable for harvest. The spread of *F. magna* into its contemporary North American distribution was dependent on opportunities for infected cervids enter new regions, either via natural migration or via human-related translocation. At present, North American populations of *F. magna* are separated in detached pockets across the continent, in five enzootic regions that possess diverse ecological conditions (Pybus 2001).

The northern Pacific coast (*NPC*) is westernmost enzootic region of coastal Canadian province British Columbia (BC), and US states Oregon (OR) and Washington (WA) (Fig. 2.2), where diverse spectrum of naturally infected final hosts was found out. *Fascioloides magna* was detected in many definitive hosts

Table 2.1 Spectrum of localities and final hosts of natural *F. magna* infections in North America

Enzootic region[a]	CA province/US state[b]	Locality	Final host[c]	P (%)	Period of examination	References
Northern Pacific coast (NPC)	CA/British Columbia/BC	Kootenay NP	White-tailed deer	28	1984–1991	Pybus et al. (2015)
	CA/British Columbia/BC	Kootenay NP	Wapiti	n.i.	n.i.	Flook and Stenton (1969)
	CA/British Columbia/BC	Kootenay NP	Wapiti	77–100	1985–1989	Pybus et al. (2015)
	CA/British Columbia/BC	Vancouver Island	Roosevelt elk	n.i.	n.i.	Bazsalovicsová et al. (2015)
	CA/British Columbia/BC	n.i.	Black-tailed deer	n.i.	n.i.	Hadwen (1916) c.i. Pybus (2001)
	CA/British Columbia/BC	n.i.	Black-tailed deer	n.i.	n.i.	Cowan (1946)
	CA/British Columbia/BC	Kootenay NP	Mule deer	4	1984–1991	Pybus et al. (2015)
	CA/British Columbia/BC	n.i.	Moose	n.i.	n.i.	Hilton (1930) c.i. Pybus (2001)
	CA/British Columbia/BC	n.i.	Moose	n.i.	n.i.	Cowan (1951)
	CA/British Columbia/BC	Kootenay NP	Moose	63	1984–1991	Pybus et al. (2015)
	CA/British Columbia/BC	n.i.	Cattle	n.i.	n.i.	Hilton (1930) c.i. Pybus (2001)
	US/Oregon/OR	Western Oregon	Wapiti	n.i.	n.i.	Dutson et al. (1967)
	US/Oregon/OR	Salem	Black-tailed deer	n.i.	n.i.	Bazsalovicsová et al. (2015)
	US/Oregon/OR	n.i.	Sheep	34.3	n.i.	Foreyt and Hunter (1980)
	US/Washington/WA	Olympic Peninsula	Roosevelt elk	n.i.	1935–1938	Schwartz and Mitchell (1945)

(continued)

Table 2.1 (continued)

Enzootic region[a]	CA province/US state[b]	Locality	Final host[c]	P (%)	Period of examination	References
Rocky Mountain trench (RMT)	CA/Alberta/AB	n.i.	White-tailed deer	n.i.	n.i.	Swales (1935)
	CA/Alberta/AB	SW Alberta, Cypress Hills	White-tailed deer	2	1988	Pybus (1990)
	CA/Alberta/AB	Banff NP	White-tailed deer	44	1984–1991	Pybus et al. (2015)
	CA/Alberta/AB	n.i.	Wapiti	n.i.	n.i.	Swales (1935)
	CA/Alberta/AB	Banff NP	Wapiti	n.i.	n.i.	Flook and Stenton (1969)
	CA/Alberta/AB	SW Alberta	Wapiti	50	1984	Kingscote et al. (1987)
	CA/Alberta/AB	S Alberta	Wapiti	80	1990	Whiting and Tessaro (1994)
	CA/Alberta/AB	North Saskatchewan River	Wapiti	3–33	1997	Kennedy et al. (1999)
	CA/Alberta/AB	Banff NP	Wapiti	53–79	1984–1991	Pybus et al. (2015)
	CA/Alberta/AB	Banff NP	Wapiti	n.i.	n.i.	Bazsalovicsová et al. (2015)
	CA/Alberta/AB	SW Alberta, Cypress Hills	Rocky Moun. elk	29	1988	Pybus (1990)
	CA/Alberta/AB	Banff NP	Rocky Moun. elk	93	1989	Pybus et al. (1991)
	CA/Alberta/AB	SW Alberta, Cypress Hills	Mule deer	0	1988	Pybus (1990)
	CA/Alberta/AB	Banff NP	Mule deer	6	1985–1989	Pybus et al. (2015)
	CA/Alberta/AB	SW Alberta, Cypress Hills	Moose	4	1988	Pybus (1990)
	CA/Alberta/AB	Banff NP	Moose	52	1984–1991	Pybus et al. (2015)
	CA/Alberta/AB	n.i.	Cattle	n.i.	n.i.	Swales (1935)

(continued)

Table 2.1 (continued)

Enzootic region[a]	CA province/US state[b]	Locality	Final host[c]	P (%)	Period of examination	References
	CA/Alberta/AB	n.i.	Bison	n.i.	n.i.	Cameron (1923) c.i. Pybus (2001)
	CA/Alberta/AB	n.i.	Bison	n.i.	n.i.	Swales (1935)
	CA/Alberta/AB	n.i.	Yak	n.i.	n.i.	Swales (1935)
	CA/Saskatchewan/SK	Central SK, Prince Albert NP	Wapiti	n.i.	1982–1983	Wobeser et al. (1985)
	CA/Saskatchewan/SK	Central SK, Prince Albert NP	Moose	n.i.	1982–1983	Wobeser et al. (1985)
	US/Montana/MT	n.i.	White-tailed deer	n.i.	n.i.	Aiton (1938) c.i. Pybus (2001)
	US/Montana/MT	n.i.	Wapiti	n.i.	n.i.	Butler (1938) c.i. Pybus (2001)
	US/Montana/MT	Flathead, McCone, Cascade, Lincoln	Wapiti	4–100	1995–1996	Hood et al. (1997)
	US/Montana/MT	n.i.	Mule deer	n.i.	n.i.	Senger (1963)
	US/Montana/MT	SW Montana	Cattle	17.2	1989–1990	Knapp et al. (1992)
	US/Montana/MT	n.i.	Sheep	n.i.	n.i.	Hall (1914) c.i. Pybus (2001)
Great Lakes region (GLR)	CA/Manitoba/MB	SE Manitoba	Moose	n.i.	n.i.	Lankester (1974)
	CA/Manitoba/MB	n.i.	Cattle	n.i.	1912	Wobeser and Schumann (2014)
	CA/Ontario/ON	n.i.	Moose	n.i.	n.i.	Kingscote (1950)
	US/Michigan/MI	S and N Michigan	Cattle	0.4–14	1977–1981	Schillhorn van Veen (1987)
	US/Minnesota/MN	n.i.	White-tailed deer	n.i.	n.i.	Fenstermacher et al. (1943)

(continued)

Table 2.1 (continued)

Enzootic region[a]	CA province/US state[b]	Locality	Final host[c]	P (%)	Period of examination	References
	US/Minnesota/MN	Erskine, Hibbing	White-tailed deer	n.i.	n.i.	Bazsalovicsová et al. (2015)
	US/Minnesota/MN	n.i.	Moose	n.i.	n.i.	Fenstermacher (1934) c.i. Pybus (2001)
	US/Minnesota/MN	NW Minnesota	Moose	n.i.	n.i.	Murray et al. (2006)
	US/Minnesota/MN	NE Minnesota	Moose	17; 5	1972–2000	Peterson et al. (2013)
	US/Minnesota/MN	Central Minnesota	Horse	n.i.	n.i.	McClanahan et al. (2005)
	US/Minnesota/MN	n.i.	Llama	n.i.	n.i.	Conboy et al. (1988)
	US/New York/NY	n.i.	White-tailed deer	n.i.	n.i.	Stiles and Hassall (1894) c.i. Pybus (2001)
	US/North Dakota/ND	n.i.	Moose	19.6	1977–1992	Maskey (2011)
	US/North Dakota/ND	n.i.	Moose	0	2002–2003	Maskey (2011)
	US/Wisconsin/WI	n.i.	Sheep	n.i.	n.i.	Campbell and Todd (1954)
Northern Quebec and Labrador (NQL)	CA/Quebec/QC	Eastern Ungava	Caribou	n.i.	n.i.	Choquette et al. (1971)
	CA/Quebec/QC	Kuujjuaq, Tasijuaq	Muskox	n.i.	n.i.	Bazsalovicsová et al. (2015)
	CA/Labrador/NL	n.i.	Caribou	58	n.i.	Lankester and Luttich (1988)
	CA/Labrador/NL	Southcentral, coastal N Labrador	Caribou	78	2001	Pollock et al. (2009)
	CA/Labrador/NL	N Labrador at Nashaupi River	Caribou	n.i.	n.i.	Bazsalovicsová et al. (2015)

(continued)

Table 2.1 (continued)

Enzootic region[a]	CA province/US state[b]	Locality	Final host[c]	P (%)	Period of examination	References
Gulf coast, lower Mississippi, and southern Atlantic seaboard (SAS)	13 southeastern US states	n.i.	White-tailed deer	12.8	n.i.	Pursglove et al. (1977)
	US/Florida/FL	n.i.	White-tailed deer	n.i.	n.i.	Dinaburg (1939) c.i. Pybus (2001)
	US/Florida/FL	White Oak plantation	White-tailed deer	n.i.	n.i.	Bazsalovicsová et al. (2015)
	US/Georgia/GA	Wilkinson	White-tailed deer	n.i.	n.i.	Bazsalovicsová et al. (2015)
	US/Kentucky/KY	Fulton County	White-tailed deer	n.i.	1986	Lydeard et al. (1989)
	US/Louisiana/LA	Tensas NWR	White-tailed deer	n.i.	n.i.	Bazsalovicsová et al. (2015)
	US/Mississippi/MS	St. Catherine NWR	White-tailed deer	n.i.	n.i.	Bazsalovicsová et al. (2015)
	US/North Carolina/NC	Halifax County	White-tailed deer	73	1993–1994	Flowers (1996)
	US/South Carolina/SC	n.i.	White-tailed deer	n.i.	n.i.	Dinaburg (1939) c.i. Pybus (2001)
	US/South Carolina/SC	SRP, Aiken and Barnwell	White-tailed deer	30	1986	Lydeard et al. (1989)
	US/South Carolina/SC	WWC, Hampton County	White-tailed deer	25.6	1986	Lydeard et al. (1989)
	US/South Carolina/SC	25 SC Counties	White-tailed deer	11.7	2002–2006	Steele (2008)
	US/South Carolina/SC	Savannah River Site	White-tailed deer	n.i.	n.i.	Bazsalovicsová et al. (2015)

(continued)

Table 2.1 (continued)

Enzootic region[a]	CA province/US state[b]	Locality	Final host[c]	P (%)	Period of examination	References
	US/Tennessee/TN	Reelfoot NWR, Obion County	White-tailed deer	41.9	1986	Lydeard et al. (1989)
	US/Tennessee/TN	SFWMA, Shelby County	White-tailed deer	53.3	1986	Lydeard et al. (1989)
	US/Texas/TX	n.i.	White-tailed deer	n.i.	n.i.	Olsen (1949)
	US/Texas/TX	n.i.	White-tailed deer	69.7	n.i.	Foreyt and Todd (1972)
	US/Texas/TX	Southern Texas	White-tailed deer	64–84	1971–1975	Foreyt et al. (1977)
	US/Texas/TX	n.i.	Cattle	n.i.	n.i.	Francis (1891) c.i. Pybus (2001)
	US/Texas/TX	n.i.	Cattle	38.3	n.i.	Foreyt and Todd (1972)
	US/Texas/TX	n.i.	Wild boar	51.7	n.i.	Foreyt and Todd (1972)
	US/Texas/TX	San Patricio and Victoria Counties	Wild boar	69	1971–1975	Foreyt et al. (1975)
	US/Texas/TX	Dimmitt County	Wild boar	n.i.	n.i.	Schwartz et al. (1993)
	US/Texas/TX	Southern Texas	Collared peccary	1	n.i.	Samuel and Low (1970)
	US/Texas/TX	n.i.	Sheep	n.i.	n.i.	Olsen (1949)
	US/Texas/TX	n.i.	Goat	n.i.	n.i.	Olsen (1949)

CA Canada, *US* United States, *P* prevalence, *n.i.* not indicated in the respective literature, *c.i.* cited in, *NP* National Park, *SW* southwestern, *S* southern, *Rocky Moun. elk* Rocky Mountain elk, *SE* southeastern, *N* northern, *NW* northwestern, *NE* northeastern, *NWR* National Wildlife Refuge, *SRP* Savannah River Plant, *WWC* Webb Wildlife Center, *SFWMA* Shelby Forest Wildlife Management Area

[a]Order of enzootic regions follows direction from west to east and from north to south

[b]CA provinces and US states within respective enzootic region are listed alphabetically

[c]Final hosts (Latin names provided in Chap. 3) within respective CA provinces/US states are listed in order as indicated in Chap. 3

species (white-tailed deer, wapiti, Roosevelt elk, black-tailed deer and mule deer), but also in moose, cattle and sheep (see Table 2.1 and references therein). The highest prevalence was documented in wapiti (77–100 %) and moose (63 %) from Kootenay National Park (BC) (Pybus et al. 2015).

The Rocky Mountain trench (RMT) includes Canadian provinces Alberta (AB) and adjacent Saskatchewan (SK), and US state Montana (MT) (Fig. 2.2). The majority of giant liver fluke findings originated from the Banff National Park (NP) and southwestern Alberta. *Fascioloides magna* was determined in white-tailed deer, wapiti, Rocky Mountain elk, mule deer (definitive hosts), but also in moose, cattle, bison, yak and sheep (see Table 2.1 and references therein). The highest prevalence (up to 80 %) was determined in wapiti from southern Alberta and Banff NP (Whiting and Tessaro 1994; Pybus et al. 2015) and in Rocky Mountain elk in Banff NP (93 %; Pybus et al. 1991). Hood et al. (1997) detected 4–100 % prevalence of fascioloidosis in wapiti from Montana.

The Great Lakes region (*GLR*) involves US states surrounding the Great Lakes, e.g. Minnesota (MN), Wisconsin (WI), Michigan (MI) and New York (NY), and Canadian province Ontario (ON). Adjacent Canadian province Manitoba (MB) and US state North Dakota (ND) are considered to be within GLR (Fig. 2.2). Contrary to NPC and RMT, the white-tailed deer was the only definitive host in GLR region. Majority of natural *F. magna* infections were detected in moose, cattle, llama, horse (dead-end hosts) and sheep (aberrant host) (see Table 2.1 and references therein). The highest prevalence (up to 20 %) was documented during the long term surveys in moose from Minnesota (Peterson et al. 2013) and North Dakota (Maskey 2011).

The Saskatchewan and Manitoba/North Dakota are not directly located in the Rocky Mountains and Great Lakes regions, respectively. Their classification into RMT and GLR enzootic regions is due to their location neighbouring the respective regions.

Northern Quebec and Labrador (*NQL*) is the northernmost enzootic region with Canadian provinces Quebec (QC) and Labrador (NL). The dominant definitive host is caribou (Table 2.1); *F. magna* was also found in muskox (Bazsalovicsová et al. 2015). The highest prevalence (78 %) was determined in caribou from southcentral and coastal northern Labrador in 2001 (Pollock et al. 2009).

Gulf coast, lower Mississippi, and southern Atlantic seaboard (SAS) offers for giant liver fluke and its hosts suitable ecological conditions, in particular moist lowlands or swamps along major drainage systems (Pursglove et al. 1977). This enzootic region includes southeastern US states Georgia (GA), Florida (FL), Kentucky (KY), Louisiana (LA), Mississippi (MS), North Carolina (NC), South Carolina (SC), Tennessee (TN) and Texas (TX) (Fig. 2.2). The dominant definitive host is white-tailed deer (see Table 2.1 and references therein). Out of dead-end hosts, *F. magna* infection was found in cattle, collared peccary and wild boar. Infection in goat and sheep (aberrant hosts) was documented in Texas (Olsen 1949). The highest prevalence (over 64 %) was documented in white-tailed deer and wild boar from Texas (Foreyt and Todd 1972; Foreyt et al. 1975, 1977).

Genetic interrelationships among populations of giant liver fluke from all enzootic regions were studied by Bazsalovicsová et al. (2015) using short variable regions of mitochodrial *cox*1 and *nad*1 markers designed by Králová-Hromadová et al. (2008). The principal outcome was detection of two separate lineages of *F. magna* in North American continent. The western lineage was formed by individuals from RMT region (in particular Alberta) and NPC region (British Columbia and Oregon). The eastern lineage was formed by samples from GLR region (Minnesota), SAS region (Mississippi, Louisiana, South Carolina, Georgia and Florida) and NQL region (Quebec and Labrador). More details on mitochondrial markers and their application in *F. magna* studies are provided in Sect. 5.2.

2.2 Europe

As a result of popular commercialized hunting in Europe of the 19th and 20th centuries, wapiti and white-tailed deer were imported from North America to European parks, zoological gardens and reservations (Ślusarski 1955; Bojović and Halls 1984). Together with the introduction of the game animals, *F. magna* was transferred as an undesirable side-effect from the Nearctic zone to the Palearctic, where it established local populations (Erhardová-Kotrlá 1971; Pybus 2001).

Contrary to native North America, Europe represents the continent where *F. magna* was introduced along with its cervid hosts and established three permanent natural foci: (1) La Mandria Regional Park in the northern Italy (IT); (2) Czech Republic and southwestern Poland (CZ-PL); and (3) Danube floodplain forests (DFF) (Fig. 2.3). The particular European countries with confirmed natural infections of final hosts are illustrated on Fig. 2.4. Details on geographic localities, final hosts and prevalence of fascioloidosis in all European natural foci are provided in Table 2.2.

Italy (IT) Fascioloides magna was introduced to the Royal Park La Mandria (now La Mandria Regional Park) near Turin in northern Italy in 1865, as was recorded and originally described by Bassi (1875). The Savoy King Vittorio Emanuele II directed import of 60 wapiti from Wyoming (USA), of which 47 reached La Mandria Regional Park (Apostolo 1996). The Regional Park represents first stable and isolated natural focus of fascioloidosis in Europe (Fig. 2.3) with occurrence of giant liver fluke mainly in red deer (see Table 2.2 and references therein). Besides, *F. magna* was also found in introduced Rocky Mountain elk and Sambar deer (Bassi 1875 c.i. Pybus 2001). A very high prevalence (up to 100 %) was recorded in red deer in a period of 1979–1980 (Balbo et al. 1987). Fascioloidosis was determined in the Italian focus in rather broad spectrum of dead-end hosts, mainly in cattle, blue bull, horse and wild boar and in sheep and goat (aberrant hosts) (see Table 2.2 and references therein).

Czech Republic and southwestern Poland (CZ-PL) The second European focus of fascioloidosis was determined in the Czech Republic by Ullrich (1930), who published the first occurrence of *F. magna* in fallow deer in 1910. The Czech focus was established mainly in the southern and central parts of the country, where *F. magna* was found in definitive hosts, such as red deer, fallow deer, or

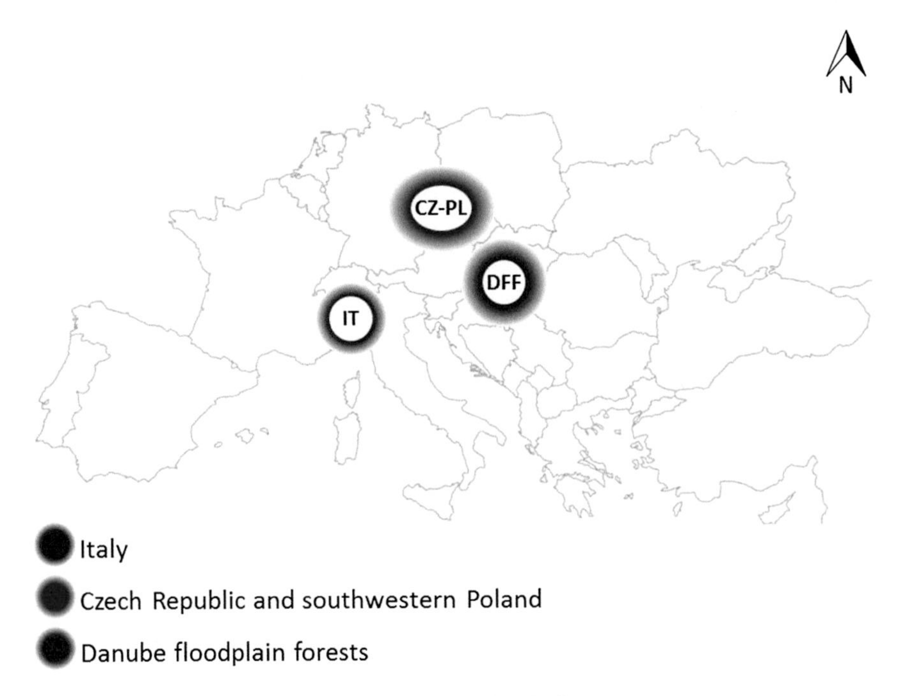

Fig. 2.3 Schematic presentation of European natural foci of *F. magna*

white-tailed deer, also in sika deer and cattle (dead-end hosts), and roe deer (aberrant host) (see Table 2.2 and references therein). The highest prevalence was detected in red deer (81–100 %) and roe deer (70–80 %; Erhardová-Kotrlá 1971), and later on in red deer and fallow deer (up to 90 %; Novobilský et al. 2007).

The Czech focus of fascioloidosis was restricted to the southern and central parts of the Czech Republic for more than 100 years. Recently, *F. magna* was documented in the northern Czech-Polish border, what indicated a possible spread of fascioloidosis from Czech Republic to Poland (Kašný et al. 2012). Indeed, parasitological survey of cervids of the Lower Silesian Wilderness and Bory Zielonogórskie (southwestern Poland) performed in the period of 2012–2013 and in 2015, respectively, confirmed presence of *F. magna* in red deer, roe deer and fallow deer (Pyziel et al. 2014; Demiaszkiewicz et al. 2015). Molecular genotypization of Polish flukes using mitochondrial markers revealed close genetic interrelationships between Czech and Polish parasites (Králová-Hromadová et al. 2015). These findings indicate that giant liver fluke has spread to Lower Silesian Wilderness from well-established Czech focus, and common name "the Czech Republic and southwestern Poland" was suggested (Králová-Hromadová et al. 2015).

It is important to mention, that the first record on fascioloidosis in Poland originated from red deer from the Lower Silesian Forest in the southwestern Poland in 1953 (Ślusarski 1955). Since then, the occurrence of *F. magna* in Poland has not been documented for almost 60 years. Was the parasite present in the southwestern

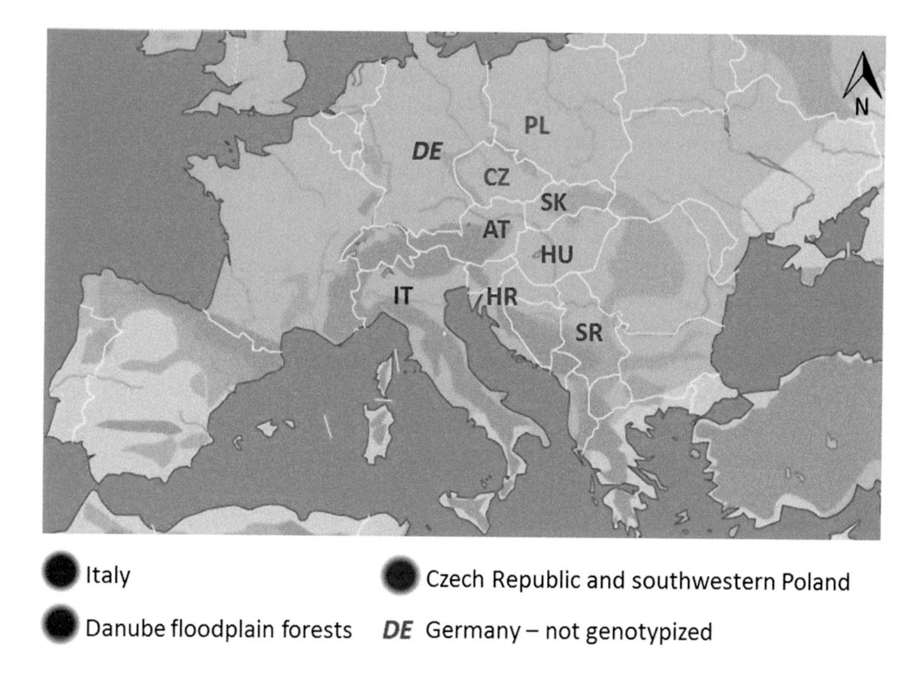

Fig. 2.4 Details on European natural foci with confirmed natural infections of *F. magna* (codes are explained in Table 2.2; map downloaded from www.johomaps.com)

Poland during the whole period from its first discovery in 1953 till present? Or does the recent detection of *F. magna* represent a new finding of the fluke in that region? Since fascioloidosis causes macroscopically visible pathological changes in livers of infected animals, it is difficult to imagine how giant liver fluke could escape from an attention of veterinarians and hunters for such a long time (Králová-Hromadová et al. 2015). More probable explanation is that *F. magna* was in 50s of the 20th century detected as a sporadic finding, but did not establish permanent focus.

The recent results of Karamon et al. (2015) confirmed that a spread of fascioloidosis is still ongoing and dynamic process which requires permanent monitoring. Giant liver fluke was found in fallow deer in southeastern region of Poland (Podkarpackie Province), neighbouring with Slovakia. The ribosomal ITS2 markers were applied for verification of the taxonomy of the fluke (Karamon et al. 2015). Since the Polish population of *F. magna* from Podkarpackie Province was not genotypized with mitochondrial markers, it can not be concluded where it originates from. It is evident that *F. magna* is expanding to novel territories, what represents potential threat for susceptible free-living and domestic ruminants.

Danube floodplain forests (*DFF*) The third European focus of fascioloidosis is Danube floodplain forests (Fig. 2.3). The first finding of *F. magna* was documented in fallow deer from game husbandry in Lower Austria in 1982 (Pfeiffer 1983). The infected animal apparently originated from the Netherlands; however, the infection did not result in establishment of the permanent focus and can be considered as a sporadic finding. The real outbreak of fascioloidosis in Danube region was documented in 90s of the 20th century, when giant liver fluke was detected in red deer in southwestern Slovakia (Rajský et al. 1994), across the Slovak-Hungarian border, in the northwestern region of Hungary named Szigetköz (Majoros and Sztojkov 1994), in Fischamend area in Austria (Winkelmayer and Prosl 2001), Baranja region of eastern Croatia (Marinculić et al. 2002) and recently in Serbia (Marinković et al. 2013). Since the first findings of the parasite in the respective countries, *F. magna* has been regularly and repeatedly detected in the Danube floodplain forests (see Table 2.2 and references therein), which represents a unique biotope located on islands of the inland delta of the Danube River. The large trans-border wetland area lacks ecological or human barriers for the movement of cervids and dispersal of the infective stages of their parasites (Králová-Hromadová et al. 2011). This natural focus of fascioloidosis is still expanding and there is a high risk that the parasite will be determined in further countries down the Danube River (e.g. Romania), or neighbouring countries, such as Bosna and Herzegovina (Sinanović et al. 2013).

Red deer represent the most frequent and dominant definitive host in Danube floodplain forests. Besides, fascioloidosis was detected also in fallow deer and roe deer (see Table 2.2 and references therein). Danube floodplain forests represent so far the only European natural focus with no documented *F. magna* infection in domestic ruminants. High prevalence (over 60 %) was repeatedly detected in red deer in southwestern Slovakia (Rajský et al. 1994, 1995, 1996, 2002; Špakulová et al. 1997), Fischamend in Austria (Winkelmayer and Prosl 2001; Ursprung and Prosl 2011), in northwestern Hungary (Giczi 2008) and Baranja region in eastern Croatia (Slavica et al. 2006).

Genetic interrelationships among *F. magna* populations from all European foci using short variable fragments of mitochondrial *cox*1 and *nad*1 revealed two independent phylogenetic lineages (Králová-Hromadová et al. 2011). The Italian population represented one phylogenetic lineage, while the second one included populations from the Czech focus and the Danube floodplain forests. Molecular data did not show any genetic relationships between flukes from Italy and other European foci. It was confirmed that *F. magna* did not spread further to Europe from Italy and this focus remained rather isolated. On the other hand, the results indicated multiple introductions of *F. magna* to Europe (Králová-Hromadová et al. 2011). More details on mitochondrial markers and their application in *F. magna* studies are provided in Sect. 5.2.

Germany (*DE*) The first record on fascioloidosis in Germany originated from 1930, when *F. magna* was found in red deer in Lower Silesia, region neighbouring Poland (Salomon 1932). This finding may be closely related with detection of *F. magna* in red deer in the Lower Silesian Forest (southwestern Poland) in 1953 (Ślusarski 1955). Similarly to the history of *F. magna* occurrence in Poland, it can be hypothesized that after introduction of giant liver fluke to Germany, the permanent focus was not established maybe due to lack of some ecological factors.

More than seven decades later, Novobilský et al. (2007) detected *F. magna* in the southwestern border of the Czech Republic indicating a threat of its possible

Table 2.2 Spectrum of localities and final hosts of natural *F. magna* infections in Europe

Natural focus[a]	Country[b]	Locality	Final host[c]	P (%)	Period of examination	References
Italy (IT)	Italy/IT	La Mandria Regional Park, northern Italy	Red deer	n.i.	**1875**	Bassi (1875) c.i. Pybus (2001)
			Red deer	51.8	1980–1983	Lanfranchi et al. (1984/85)
			Red deer	50–100	1979–1980	Balbo et al. (1987)
			Red deer	n.i.	1977–1978	Balbo et al. (1989)
			Red deer	n.i.	n.i.	Králová-Hromadová et al. (2011)
			RM elk	n.i.	**1875**	Bassi (1875) c.i. Pybus (2001)
			Fallow deer	n.i.	**1875**	Bassi (1875) c.i. Pybus (2001)
			Sambar deer	n.i.	**1875**	Bassi (1875) c.i. Pybus (2001)
			Cattle	3.4	1980–1983	Lanfranchi et al. (1984/85)
			Cattle	3.7	1979–1980	Balbo et al. (1987)
			Blue bull	n.i.	**1875**	Bassi (1875) c.i. Pybus (2001)
			Horse	5.7	1979–1980	Balbo et al. (1987)
			Wild boar	n.i.	1979–1980	Balbo et al. (1987)
			Wild boar	n.i.	1979	Balbo et al. (1989)
			Sheep	n.i.	**1875**	Bassi (1875) c.i. Pybus (2001)
			Goat	n.i.	**1875**	Bassi (1875) c.i. Pybus (2001)
Czech Republic and southwestern Poland (CZ-PL)	Czech Republic/CZ	Southern Bohemia	WT deer	n.i.	1966–1967	Erhardová-Kotrlá (1971)
		Southern Bohemia	Red deer	81–100	n.i.	Erhardová-Kotrlá (1971)
		n.i.	Red deer	n.i.	n.i.	Kolář (1978)
		Western, southern, central Bohemia	Red deer	4–95	2003–2005	Novobilský et al. (2007)

(continued)

Table 2.2 (continued)

Natural focus[a]	Country[b]	Locality	Final host[c]	P (%)	Period of examination	References
		Southern and central Bohemia	Red deer	n.i.	n.i.	Králová-Hromadová et al. (2011)
		Southern Bohemia	Fallow deer	n.i.	**1910**	Ullrich (1930)
		Southern Bohemia	Fallow deer	21.6–31.9	n.i.	Erhardová-Kotrlá (1971)
		Southern Bohemia	Fallow deer	15.8	2002–2003	Chroust and Chroustová (2004)
		Western, southern, central Bohemia	Fallow deer	28–90	2003–2005	Novobilský et al. (2007)
		Southern Bohemia	Sika deer	4	n.i.	Erhardová-Kotrlá (1971)
		Southern Bohemia	Cattle	n.i.	1965	Záhoř et al. (1966)
		Southern Bohemia	Cattle	9.1–21.1	1976–1977	Chroustová et al. (1980)
		Southern Bohemia	Cattle	n.i.	2011–2012	Leontovyč et al. (2014)
		Southern Bohemia	Roe deer	n.i.	n.i.	Záhoř (1965)
		Southern Bohemia	Roe deer	70–80	n.i.	Erhardová-Kotrlá (1971)
		Southern Bohemia	Roe deer	9.1	2002–2003	Chroust and Chroustová (2004)
	Poland/PL	Lower Silesian Wilderness	Red deer	n.i.	**1953**	Ślusarski (1955)
		Lower Silesian Wilderness	Red deer	n.i.	2012–2013	Pyziel et al. (2014); Králová-Hromadová et al. (2015)
		Bory Zielonogórskie	Red deer	n.i.	2015	Demiaszkiewicz et al. (2015)
		Bory Zielonogórskie	Fallow deer	n.i.	2015	Demiaszkiewicz et al. (2015)
		Podkarpackie Province*	Fallow deer	n.i.	2015	Karamon et al. (2015)
		Bory Zielonogórskie	Roe deer	n.i.	2015	Demiaszkiewicz et al. (2015)
Danube floodplain forests (DFF)	Austria/AT	Game husbandry, Lower Austria	Fallow deer	n.i.	**1982**	Pfeiffer (1983)
		Fischamend	Red deer	66.7	2000	Winkelmayer and Prosl (2001)
		Fischamend	Red deer	15.8	2000–2005	Ursprung et al. (2006)

(continued)

Table 2.2 (continued)

Natural focus[a]	Country[b]	Locality	Final host[c]	P (%)	Period of examination	References
		Fischamend	Red deer	13–100	2000–2010	Ursprung and Prosl (2011)
		Fischamend	Roe deer	n.i.	2000	Winkelmayer and Prosl (2001)
		Fischamend	Roe deer	n.i.	2000–2005	Ursprung et al. (2006)
	Hungary/HU	NW Hungary, Szigetköz	Red deer	n.i.	**1991**	Majoros and Sztojkov (1994)
		NW Hungary, Szigetköz	Red deer	21.1–65.1	1999–2006	Giczi (2008)
		NW Hungary, Szigetköz	Red deer	n.i.	n.i.	Králová-Hromadová et al. (2011)
		NW Hungary, Szigetköz	Roe deer	3.7	1999–2006	Giczi (2008)
	Slovakia/SK	SW Slovakia, Gabčíkovo	Red deer	100	**1993**	Rajský et al. (1994)
		SW Slovakia	Red deer	70–80	1991–1995	Rajský et al. (1995)
		SW Slovakia, Gabčíkovo	Red deer	90	n.i.	Rajský et al. (1996)
		SW Slovakia	Red deer	70	n.i.	Špakulová et al. (1997)
		SW Slovakia, Dunajská Streda	Red deer	91.3	1993–2001	Rajský et al. (2002)
		SW Slovakia	Red deer	17.39	2005	Rajský et al. (2006)
		SW Slovakia	Red deer	n.i.	n.i.	Králová-Hromadová et al. (2011)
		SW Slovakia, Dunajská Streda	Roe deer	60	1993–2001	Rajský et al. (2002)
		SW Slovakia	Roe deer	n.i.	2005	Rajský et al. (2006)

(continued)

Table 2.2 (continued)

Natural focus[a]	Country[b]	Locality	Final host[c]	P (%)	Period of examination	References
	Croatia/HR	E Croatia, Baranja region	Red deer	n.i.	**2000**	Marinculić et al. (2002)
		E Croatia, Baranja region	Red deer	54.1	2000–2001	Janicki et al. (2005)
		E Croatia, Baranja region	Red deer	20–80	2001–2003	Slavica et al. (2006)
		E Croatia, Baranja region	Red deer	53.3	2002–2003	Rajković-Janje et al. (2008)
		Central, littoral, W Croatia	Red deer	4.05	2002–2003	Rajković-Janje et al. (2008)
		E Croatia, Baranja region	Red deer	0–48	2001–2004	Florijančić et al. (2010)
		E Croatia, Baranja region	Red deer	n.i.	n.i.	Králová-Hromadová et al. (2011)
		E Croatia	Red deer	57.4	2006–2008	Severin et al. (2012)
	Serbia/SR	Northern Serbia	Fallow deer	52.2	n.i.	Marinković et al. (2013)
	Germany/DE*	Lower Silesia	Red deer	n.i.	**1930**	Salomon (1932)
		Northeastern Bavaria	Red deer	70	2011–2012	Rehbein et al. (2012)
		Northeastern Bavaria	Red deer	4.9	2012–2013	Plötz et al. (2015)
		Northeastern Bavaria	Fallow deer	10.2	2012–2013	Plötz et al. (2015)
		Northeastern Bavaria	Sika deer	37.5	2011–2012	Rehbein et al. (2012)
		Northeastern Bavaria	Roe deer	20	2011–2012	Rehbein et al. (2012)

n.i. not indicated in the respective literature, *c.i.* cited in, *P* prevalence, *RM elk* Rocky Mountain elk, *WT deer* white-tailed deer, *NW* northwestern, *SW* southwestern, *E* eastern, *W* western

*German population and Polish population from Podkarpackie Province have not been genotypized yet, therefore, the exact natural focus is not known, *years in bold* indicate the first finding of *F. magna* in the respective country

[a]Chronological order of natural foci

[b]Chronological order of European countries within respective focus

[c]Species of final hosts (Latin names provided in Chap. 3) within respective countries are listed in order as indicated in Chap. 3

spread into Germany. This suspicion was proved to be correct, since fascioloidosis in red deer, roe deer, sika deer, and fallow deer was reported in northeastern Bavaria (Rehbein et al. 2012; Plötz et al. 2015). *Fascioloides magna* specimens from Germany have not been genotypized, hence their exact origin is not known. Giant liver fluke in Bavaria very probably originates from the Czech focus; however, molecular characterization and studies on population genetics of the parasite in this region need to be performed in order to determine its origin.

2.3 Sporadic Reports

In Europe, the occasional finding of fascioloidosis was reported from Spain (Almarza 1935). In a global scale, fascioloidosis was documented in Brahman heifer in the South Africa (Boomker and Dale-Kuys 1977), ox in Australia (Arundel and Hamir 1982), and from wapiti in Cuba (Lorenzo et al. 1989). The giant liver fluke was imported to these localities mainly from North America. Probably due to unsuitable environmental and climate conditions, the life cycle of the parasite could not be completed and permanent foci were not established (Pybus 2001).

References

Aiton JF (1938) Enlarged spleen in white-tailed deer at Glacier National Park. Trans North Am Wildl Conf 3:890–892. Cited in Pybus MJ (2001) Liver flukes. In: Samuel WM, Pybus MJ, Kocan AA (eds) Parasitic diseases of wild mammals, 2nd edn. Iowa State University Press, Ames

Almarza N (1935) The liver fluke in sheep. Description of new species. Infect Dis Hyg Pets 47:195–202 (in German)

Apostolo C (1996) The naturalistic aspects: flora, fauna and the environment. In: Lupo M, Paglieri M, Apostolo C, Vaccarino E, Debernardi M (eds) La Mandria Storia e natura del Parco. Edizioni Eda, Torino

Arundel JH, Hamir AN (1982) *Fascioloides magna* in cattle. Aust Vet J 58:35–36

Balbo T, Lanfranchi P, Rossi L, Meneguz PG (1987) Health management of a red deer population infected by *Fascioloides magna* (Bassi, 1875) Ward, 1917. Ann Fac Med Vet Torino 32:1–13

Balbo T, Rossi P, Meneguz PG (1989) Integrated control of *Fascioloides magna* infection in northern Italy. Parassitologia 31:137–144

Bassi R (1875) Sulla cachessia ittero-verminosa, o marciaia, causta dei Cervi, causata dal *Distomum magnum*. Il Medico Veterinario 4:497–515. Cited in Pybus MJ (2001) Liver flukes. In: Samuel WM, Pybus MJ, Kocan AA (eds) Parasitic diseases of wild mammals, 2nd edn. Iowa State University Press, Ames

Bazsalovicsová E, Králová-Hromadová I, Štefka J, Minárik G, Bokorová S, Pybus M (2015) Genetic interrelationships of North American populations of giant liver fluke *Fascioloides magna*. Parasit Vectors 8:1–15. doi:10.1186/s13071-015-0895-1

Bojović D, Halls LK (1984) Central Europe. In: Halls LK (ed) White-tailed deer ecology and management. Stackpole Books, Harrisburg

Boomker J, Dale-Kuys JC (1977) First report of *Fascioloides magna* (Bassi, 1875) in South Africa. Onderstepoort J Vet Res 44:49–52

Butler WJ (1938) Wild animal disease investigation. Montana Livestock Sanitary Board 1:18–19. Cited in Pybus MJ (2001) Liver flukes. In: Samuel WM, Pybus MJ, Kocan AA (eds) Parasitic diseases of wild mammals, 2nd edn. Iowa State University Press, Ames

Cameron AE (1923) Notes on buffalo: anatomy, pathological conditions, and parasites. Brit Vet J 79:331–336. Cited in Pybus MJ (2001) Liver flukes. In: Samuel WM, Pybus MJ, Kocan AA (eds) Parasitic diseases of wild mammals, 2nd edn. Iowa State University Press, Ames

Campbell WC, Todd AC (1954) Natural infections of *Fascioloides magna* in Wisconsin sheep. J Parasitol 40:100. Cited in Pybus MJ (2001) Liver flukes. In: Samuel WM, Pybus MJ, Kocan AA (eds) Parasitic diseases of wild mammals, 2nd edn. Iowa State University Press, Ames

Choquette LP, Gibson GG, Simard B (1971) *Fascioloides magna* (Bassi, 1875) Ward, 1917 (Trematoda) in woodland caribou, *Rangifer tarandus caribou* (Gmelin), of northeastern Quebec, and its distribution in wild ungulates in Canada. Can J Zool 49:280–281

Chroust K, Chroustová E (2004) Motolice obrovská (*Fascioloides magna*) u spárkaté zvěře v jihočeských lokalitách. Veterinářství 54:296–304 (in Czech)

Chroustová E, Hůlka J, Jaroš J (1980) Prevence a terapie fascioloidózy skotu bithionolsulfoxidem. Vet Med (Praha) 25:557–563 (in Czech)

Conboy GA, O'Brien TD, Stevens DL (1988) A natural infection of *Fascioloides magna* in a llama (*Lama glama*). J Parasitol 74:345–346

Cowan IM (1946) Parasites, diseases, injuries, and anomalies of the Columbian black-tailed deer, *Odocoileus hemionus columbianus* (Richardson), in British Columbia. Can J Res 24:71–103

Cowan IM (1951) The diseases and parasites of big game mammals of western Canada. Proc Ann Game Convention 5:37–64

Demiaszkiewicz AW, Kuligowska I, Pyziel AM, Lachowicz J, Kowalczyk R (2015) Extension of occurrence area of the American fluke *Fascioloides magna* in south-western Poland. Ann Parasitol 61:93–96

Dinaburg AG (1939) Helminth parasites collected from deer, *Odocoileus virginianus* in Florida. Proc Helminthol Soc Wash 6:102–104. Cited in Pybus MJ (2001) Liver flukes. In: Samuel WM, Pybus MJ, Kocan AA (eds) Parasitic diseases of wild mammals, 2nd edn. Iowa State University Press, Ames

Dutson VJ, Shaw JN, Knapp SE (1967) Epizootiologic factors of *Fascioloides magna* (Trematoda) in Oregon and southern Washington. Am J Vet Res 28:853–860

Erhardová-Kotrlá B (1971) The occurrence of *Fascioloides magna* (Bassi, 1875) in Czechoslovakia. Czechoslovak Academy of Sciences, Prague

Fenstermacher R (1934) Diseases affecting moose. Alumni Q 22:81–94. Cited in Pybus MJ (2001) Liver flukes. In: Samuel WM, Pybus MJ, Kocan AA (eds) Parasitic diseases of wild mammals, 2nd edn. Iowa State University Press, Ames

Fenstermacher R, Olsen OW, Pomeroy BS (1943) Some diseases of white-tailed deer in Minnesota. Cornell Vet 33:323–332

Flook DR, Stenton JE (1969) Incidence and abundance of certain parasites in wapiti in the national parks of the Canadian Rockies. Can J Zool 47:795–803. doi:10.1139/z69-138

Florijančić T, Ozimec S, Opačak A, Bošković I, Jelkić D, Marinculić A, Janicki Z (2010) Importance of the Danube River in spreading the infection of red deer with *Fascioloides magna* in eastern Croatia. Paper presented at 38th IAD conference, Dresden, Germany, 22–25 June 2010

Flowers J (1996) Notes on the life history of *Fascioloides magna* (Trematoda) in North Carolina. J Elisha Mitch Sci S 112:115–118

Foreyt WJ, Hunter RL (1980) Clinical *Fascioloides magna* infection in sheep in Oregon on pasture shared by Columbian white-tailed deer. Am J Vet Res 41:1531–1532

Foreyt WJ, Todd AC (1972) The occurrence of *Fascioloides magna* and *Fasciola hepatica* together in the livers of naturally infected cattle in South Texas, and the incidence of the flukes in cattle, white-tailed deer, and feral hogs. J Parasitol 58:1010–1011

Foreyt WJ, Todd AC, Foreyt K (1975) *Fascioloides magna* (Bassi, 1875) in feral swine from southern Texas. J Wildl Dis 11:554–559. doi:10.7589/0090-3558-11.4.554

Foreyt WJ, Samuel WM, Todd AC (1977) *Fascioloides magna* in white-tailed deer (*Odocoileus virginianus*): observation of the pairing tendency. J Parasitol 63:1050–1052. doi:10.2307/3279843

Francis M (1891) Liver flukes. Tex AES Bull 18:123–136. Cited in Pybus MJ (2001) Liver flukes. In: Samuel WM, Pybus MJ, Kocan AA (eds) Parasitic diseases of wild mammals, 2nd edn. Iowa State University Press, Ames

Giczi E (2008) *Fascioloides magna* (Bassi, 1875) infection of Hungarian red deer and roe deer stock and the possibility of protection. Dissertation, University of West Hungary

Hadwen S (1916) A new host for *Fasciola magna*, Bassi, together with observation on the distribution of *Fasciola hepatica*, L. in Canada. J Am Vet Med Assoc 49:511–515. Cited in Pybus MJ (2001) Liver flukes. In: Samuel WM, Pybus MJ, Kocan AA (eds) Parasitic diseases of wild mammals, 2nd edn. Iowa State University Press, Ames

Hall MC (1914) Society proceedings of the Helminthological Society of Washington. J Parasitol 1:106. Cited in Pybus MJ (2001) Liver flukes. In: Samuel WM, Pybus MJ, Kocan AA (eds) Parasitic diseases of wild mammals, 2nd edn. Iowa State University Press, Ames

Hilton G (1930) Report of the Veterinary Director General, Department of Agriculture, Ottawa, Canada. Cited in Pybus MJ (2001) Liver flukes. In: Samuel WM, Pybus MJ, Kocan AA (eds) Parasitic diseases of wild mammals, 2nd edn. Iowa State University Press, Ames

Hood BR, Rognlie MC, Knapp SE (1997) Fascioloidiasis in game-ranched elk from Montana. J Wildl Dis 33:882–885. doi:10.7589/0090-3558-33.4.882

Janicki Z, Konjević D, Severin K (2005) Monitoring and treatment of *Fascioloides magna* in semi-farm red deer husbandry in Croatia. Vet Res Commun 29:83–88. doi:10.1007/s11259-005-0027-z

Karamon J, Larska M, Jasik A, Sell B (2015) First report of the giant liver fluke (*Fascioloides magna*) infection in farmed fallow deer (*Dama dama*) in Poland—pathomorphological changes and molecular identification. Bull Vet Inst Pulawy 59:339–344. doi:10.1515/bvip-2015-0050

Kašný M, Beran L, Siegelová V, Siegel T, Leontovyč R, Beránková K, Pankrác J, Košťáková M, Horák P (2012) Geographical distribution of the giant liver fluke (*Fascioloides magna*) in the Czech Republic and potential risk of its further spread. Vet Med 57:101–109

Kennedy MJ, Acorn RC, Moraiko DT (1999) Survey of *Fascioloides magna* in farmed wapiti in Alberta. Can Vet J 40:252–254

Kingscote AA (1950) Liver rot (Fascioloidiasis) in ruminants. Can J Comp Med Vet Sci 14:203–208

Kingscote BF, Yates WDG, Tiffin GB (1987) Diseases of wapiti utilizing cattle range in southwestern Alberta. J Wildl Dis 23:86–91

Knapp SE, Dunkel AM, Han K, Zimmerman LA (1992) Epizootiology of fascioliasis in Montana. Vet Parasitol 42:241–246

Kolář Z (1978) Příspěvek k léčbě fascioloidózy u jelení zvěře. Veterinářství 28:276–277 (in Czech)

Králová-Hromadová I, Špakulová M, Horáčková E, Turčeková L, Novobilský A, Beck R, Koudela B, Marinculić A, Rajský D, Pybus M (2008) Sequence analysis of ribosomal and mitochondrial genes of the giant liver fluke *Fascioloides magna* (Trematoda: Fasciolidae): intraspecific variation and differentiation from *Fasciola hepatica*. J Parasitol 94:58–67. doi:10.1645/GE-1324.1

Králová-Hromadová I, Bazsalovicsová E, Štefka J, Špakulová M, Vávrová S, Szemes T, Tkach V, Trudgett A, Pybus M (2011) Multiple origins of European populations of the giant liver fluke *Fascioloides magna* (Trematoda: Fasciolidae), a liver parasite of ruminants. Int J Parasitol 41:373–383. doi:10.1016/j.ijpara.2010.10.010

Králová-Hromadová I, Bazsalovicsová E, Demiaszkiewicz A (2015) Molecular characterization of *Fascioloides magna* (Trematoda: Fascioloidae) from south-western Poland based on mitochondrial markers. Acta Parasitol 60:544–547. doi:10.1515/ap-2015-0077

Lanfranchi P, Tolari F, Forletta R, Meneguz PG, Rossi L (1984/85) The red deer as reservoir of parasitic and infectious pathogens for cattle. Ann Fac Med Vet Torino 30:1–17

Lankester MW (1974) *Parelaphostrongylus tenuis* (Nematoda) and *Fascioloides magna* (Trematoda) in moose of southeastern Manitoba. Can J Zool 52:235–239

Lankester MW, Luttich S (1988) *Fascioloides magna* (Trematoda) in woodland caribou (*Rangifer tarandus caribou*) of the George River herd, Labrador. Can J Zool 66:475–479. doi:10.1139/z88-067

Leontovyč R, Košťáková M, Siegelová V, Melounová K, Pankrác J, Vrbová K, Horák P, Kašný M (2014) Highland cattle and *Radix labiata*, the hosts of *Fascioloides magna*. BMC Vet Res 10:1–7. doi:10.1186/1746-6148-10-41

Lorenzo M, Ramirez P, Mendez M, Alonso M, Ramos R (1989) Reporte de *Fascioloides magna*, Bassi, 1875, parasitando un wápiti (*Cervus canadensis*) en Cuba. Rev Cubana Cien Veterinarias 20:263–266

Lydeard C, Mulvey M, Aho JM, Kennedy PK (1989) Genetic variability among natural populations of the liver fluke *Fascioloides magna* in white-tailed deer, *Odocoileus virginianus*. Can J Zool 67:2021–2025. doi:10.1139/z89-287

Majoros G, Sztojkov V (1994) Appearance of the large American liver fluke *Fascioloides magna* (Bassi, 1875) (Trematoda: Fasciolata) in Hungary. Parasit Hung 27:27–38

Marinculić A, Džakula N, Janicki Z, Hardy Z, Lučinger S, Živičnjak T (2002) Appearance of American liver fluke (*Fascioloides magna*, Bassi, 1875) in Croatia—a case report. Vet Arhiv 72:319–325

Marinković D, Kukolj V, Aleksić-Kovačević S, Jovanović M, Knežević M (2013) The role of hepatic myofibroblasts in liver cirrhosis in fallow deer (*Dama dama*) naturally infected with giant liver fluke (*Fascioloides magna*). BMC Vet Res 9:45. doi:10.1186/1746-6148-9-45

Maskey JJ (2011) Giant liver fluke in North Dakota moose. Alces 47:1–7

McClanahan SL, Stromberg BE, Hayden DW, Averbeck GA, Wilson JH (2005) Natural infection of a horse with *Fascioloides magna*. J Vet Diagn Invest 17:382–385. doi:10.1177/104063870501700415

Murray DL, Cox EW, Ballard WB, Whitlaw HA, Lenarz MS, Custer TW, Barnett T, Fuller TK (2006) Pathogens, nutritional deficiency, and climate influences on a declining moose population. Wildl Monogr 166:1–30. doi:10.2193/0084-0173(2006)166

Novobilský A, Horáčková E, Hirtová L, Modrý D, Koudela B (2007) The giant liver fluke *Fascioloides magna* (Bassi 1875) in cervids in the Czech Republic and potential of its spreading to Germany. Parasitol Res 100:549–553. doi:10.1007/s00436-006-0299-4

Olsen OW (1949) White-tailed deer as a reservoir of the large American liver fluke. Vet Med 44:26–30

Peterson WJ, Lankester MW, Kie JG, Bowyer RT (2013) Geospatial analysis of giant liver flukes among moose: effects of white-tailed deer. Acta Theriologica 58:359–365. doi:10.1007/s13364-013-0130-4

Pfeiffer H (1983) *Fascioloides magna*: Erster Fund in Österreich. Wien Tierarztl Monat 70:168–170 (in German)

Plötz C, Rehbein S, Bamler H, Reindl H, Pfister K, Scheuerle MC (2015) *Fascioloides magna*—epizootiology in a deer farm in Germany. Berl Munch Tierarztl Wochenschr 128:177–182. doi:10.2376/0005-9366-128-177

Pollock B, Penashue B, McBurney S, Vanleeuwen J, Daoust PY, Burgess NM, Tasker AR (2009) Liver parasites and body condition in relation to environmental contaminants in caribou (*Rangifer tarandus*) from Labrador, Canada. Arctic 62:1–12

Pursglove SR, Prestwood AK, Ridgeway TR, Hayes FA (1977) *Fascioloides magna* infection in white-tailed deer of southeastern United States. J Am Vet Med Assoc 171:936–938

Pybus MJ (1990) Survey of hepatic and pulmonary helminths of wild cervids in Alberta, Canada. J Wildl Dis 26:453–459. doi:10.7589/0090-3558-26.4.453

Pybus MJ (2001) Liver flukes. In: Samuel WM, Pybus MJ, Kocan AA (eds) Parasitic diseases of wild mammals, 2nd edn. Iowa State University Press, Ames

Pybus MJ, Onderka DK, Cool N (1991) Efficacy of triclabendazole against natural infections of *Fascioloides magna* in wapiti. J Wildl Dis 27:599–605. doi:10.7589/0090-3558-27.4.599

Pybus MJ, Butterworth EW, Woods JG (2015) An expanding population of the giant liver fluke (*Fascioloides magna*) in elk (*Cervus canadensis*) and other ungulates in Canada. J Wildl Dis 51:431–445. doi:10.7589/2014-09-235

Pyziel AM, Demiaszkiewicz AW, Kuligowska I (2014) Molecular identification of *Fascioloides magna* (Bassi, 1875) from red deer from south-western Poland (Lower Silesian Wilderness) on the basis of internal transcribed spacer 2 (ITS-2). Pol J Vet Sci 17:523–525. doi:10.2478/pjvs-2014-0077

Rajković-Janje R, Bosnić S, Rimac D, Gojmerac T (2008) The prevalence of American liver fluke *Fascioloides magna* (Bassi 1875) in red deer from Croatian hunting grounds. Eur J Wildl Res 54:525–528. doi:10.1007/s10344-007-0163-6

Rajský D, Patus A, Bukovjan K (1994) Prvý nález *Fascioloides magna* Bassi, 1875 na Slovensku. Slov Vet Čas 19:29–30 (in Slovak)

Rajský D, Patus A, Bielik J (1995) Záchyt cicavice obrovskej (*Fascioloides magna* Bassi, 1875) v rámci monitoringu bioty v oblasti Vodného diela Gabčíkovo. In: Kontrišová O, Kočík K, Bublinec E (eds). Zborník referátov, Monitorovanie a hodnotenie stavu životného prostredia. Technická univerzita, Zvolen (in Slovak)

Rajský D, Patus A, Špakulová M (1996) Rozšírenie cicavice obrovskej (*Fascioloides magna* Bassi, 1875) v jelenej chovateľskej oblasti J–I Podunajská. In: Zborník referátov a príspevkov medzinárodnej konferencie 1996. Výskumný ústav živočíšnej výroby, Nitra (in Slovak)

Rajský D, Čorba J, Várady M, Špakulová M, Cabadaj R (2002) Control of fascioloidosis (*Fascioloides magna* Bassi, 1875) in red deer and roe deer. Helminthologia 39:67–70

Rajský D, Dubinský P, Krupicer I, Sabo R, Sokol J (2006) Výskyt propagačných štádií *Fascioloides magna* a iných helmintov vo fekáliách jelenej zveri z okresov hraničiacich s riekami Dunaj a Morava. Slov Vet Čas 31:177–180 (in Slovak)

Rehbein S, Hamel D, Reindl H, Visser M, Pfister K (2012) *Fascioloides magna* and *Ashworthius sidemi*—two new parasites in wild ungulates in Germany. In: Program and abstracts of the XI European multicolloquium of parasitology (EMOP XI), Cluj-Napoca, Romania, 25–29 July 2012

Salomon S (1932) *Fascioloides magna* bei deutschem Rotwild. Berl Tierärztl Wochenschr 48:627–628 (in German)

Samuel WM, Low WA (1970) Parasites of the collared peccary from Texas. J Wildl Dis 6:16–23

Schillhorn van Veen TW (1987) Prevalence of *Fascioloides magna* in cattle and deer in Michigan. J Am Vet Med Assoc 191:547–548

Schwartz JE, Mitchell GE (1945) The Roosevelt elk on the Olympic Peninsula, Washington. J Wildl Manage 9:295–319. doi:10.2307/3796372

Schwartz WL, Lawhorn DB, Montgomery E (1993) *Fascioloides magna* in a feral pig. Swine Health Prod 1:27

Senger CM (1963) Some parasites of Montana deer. Montana Wildl Autumn 5–13

Severin K, Mašek T, Janicki Z, Konjević D, Slavica A, Marinculić A, Martinković F, Vengušt G, Džaja P (2012) Liver enzymes and blood metabolites in a population of free-ranging red deer (*Cervus elaphus*) naturally infected with *Fascioloides magna*. J Helminthol 86:190–196. doi:10.1017/S0022149X1100023X

Sinanović N, Omeragić J, Zuko A, Jažić A (2013) Impact of deer migration on spread of giant American fluke (*Fascioloides magna*) in Bosna a Herzegovina. Veterinaria 62:213–222

Slavica A, Florijančić T, Janicki Z, Konjević D, Severin K, Marinculić A, Pintur K (2006) Treatment of fascioloidosis (*Fascioloides magna*, Bassi 1875) in free ranging and captive red deer (*Cervus elaphus* L.) at eastern Croatia. Vet Arhiv 76:9–18

Ślusarski W (1955) Studia nad europejskimi przedstawicielami przywry *Fasciola magna* (Bassi, 1875) Stiles, 1894. Acta Parasitol Pol 3:1–59 (in Polish)

Steele E (2008) Prevalence of the large liver fluke, *Fascioloides magna*, in the white-tailed deer in South Carolina. Paper presented at the 4th annual USC upstate research symposium, University of South Carolina, Spartanburg, 11 April 2008

Stiles CW, Hassall A (1894) The anatomy of the large American fluke (*Fasciola magna*) and a comparison with other species of the genus *Fasciola*. J Comp Med Vet Arch 15:161–178, 225–

243, 299–313, 407–417, 457–462. Cited in Pybus MJ (2001) Liver flukes. In: Samuel WM, Pybus MJ, Kocan AA (eds) Parasitic diseases of wild mammals, 2nd edn. Iowa State University Press, Ames

Swales WE (1935) The life cycle of *Fascioloides magna* (Bassi, 1875), the large liver fluke of ruminants, in Canada. Can J Res 12:177–215. doi:10.1139/cjr35-015

Špakulová M, Čorba J, Varády M, Rajský D (1997) Bionomy, distribution and importance of giant liver fluke (*Fascioloides magna*), an important parasite of free-living ruminants. Vet Med 42:139–148

Ullrich K (1930) Über das Vorkommen von seltenen oder wenig bekannten Parasiten der Säugetiere und Vögel in Böhmen und Mähren. Prag Arch Tiermed 10:19–43 (in German)

Ursprung J, Prosl H (2011) Vorkommen und Bekämpfung des Amerikanischen Riesenleberegels (*Fascioloides magna*) in den österreichischen Donauauen östlich von Wien 2000–2010. Wien Tierarztl Monat 98:275–284 (in German)

Ursprung J, Joachim A, Prosl H (2006) Epidemiology and control of the giant liver fluke, *Fascioloides magna*, in a population of wild ungulates in the Danubian wetlands east of Vienna. Berl Munch Tierarztl Wochenschr 119:316–323 (in German)

Whiting TL, Tessaro SV (1994) An abattoir study of tuberculosis in a herd of farmed elk. Can Vet J 35:497–501

Winkelmayer R, Prosl H (2001) Riesenleberegel—jetzt auch bei uns? Österreichisches Weidwerk 3:42–44 (in German)

Wobeser BK, Schumann F (2014) *Fascioloides magna* infection causing fatal pulmonary hemorrhage in a steer. Can Vet J 55:1093–1095

Wobeser G, Gajadhar AA, Hunt HM (1985) *Fascioloides magna*: occurrence in Saskatchewan and distribution in Canada. Can Vet J 26:241–244

Záhoř Z (1965) Výskyt velké motolice (*Fascioloides magna* Bassi, 1875) u srnčí zvěře. Veterinářství 15:329–324 (in Czech)

Záhoř Z, Prokeš C, Vítovec L (1966) Nález vajíček motolice *Fascioloides magna* (Bassi, 1875) a fascioloidózních změn v játrech skotu. Vet Med 39:397–404 (in Czech)

Chapter 3
Final Hosts of *Fascioloides magna*

Abstract *Fascioloides magna* parasitizes in a broad spectrum of final hosts, mainly free living and domestic ruminants. Final hosts of giant liver fluke are divided into three types (definitive, dead-end and aberrant) according to interrelationships between the parasite and the host, the ability of fluke to reach maturity and produce eggs, pathological changes within the host organism, and the potential to release eggs of *F. magna* into external environment. Definitive hosts contribute significantly to further spread of propagative stages of *F. magna* into the environment. Mature flukes localized in thin-walled pseudocysts or fibrous capsules in the liver parenchyma can produce eggs and release them into the host's small intestine through the bile system. Definitive hosts tolerate fascioloidosis rather well, and infection is very often subclinical. In dead-end hosts, giant liver fluke can reach the liver but parasite matures very rarely. Only few eggs are produced and they are not released into the bile system. In aberrant hosts, giant liver fluke can not successfully complete the migration; parasite may occasionaly move up to the liver but formation of pseudocysts is not successful. Such hosts may often die due to tissue damage, which is associated with migration of immature flukes through peritoneal, thoracic or abdominal cavities.

Keywords Giant liver fluke · Final host · Definitive host · Dead-end host · Aberrant host · Host-parasite interrelationship · Experimental infection · Natural infection

3.1 Naturally Infected Final Hosts

Natural infections of *F. magna* occur primarily in representatives of the families Cervidae and Bovidae. According to Pybus (2001), there are three basic categories of final hosts: definitive, dead-end and aberrant. Different terminology has been applied throughout the literature in order to designate various types of final hosts of giant liver fluke. In particular "obligate", "specific", "typical" and "normal" have sometimes been used to determine definitive hosts, while terms "non-specific",

I. Králová-Hromadová et al., *The Giant Liver Fluke, Fascioloides magna: Past, Present and Future Research*, SpringerBriefs in Animal Sciences,
DOI 10.1007/978-3-319-29508-4_3

"unspecific", "atypical" and "abnormal" are used to describe dead-end and aberrant hosts. Since the terminology described by Pybus (2001) takes into consideration host-parasite relationships, pathological changes within the final hosts, reproduction and further spread of the parasite, we accept and apply this terminology throughout the publication.

3.1.1 Definitive Hosts

Definitive hosts are characterized by maturation of *F. magna* flukes in thin-walled pseudocysts (fibrous capsules) in the liver parenchyma. Mature flukes produce eggs which are released into the host's small intestine through the bile system. Thus, definitive hosts contribute significantly to further spread of propagative stages (eggs) of *F. magna* into the external environment. All definitive hosts are members of the family Cervidae, and except for the red deer and fallow deer, they are primarily "New World" cervids. From veterinary point of view, infection in this type of hosts is very often subclinical and fascioloidosis is rather well tolerated. The following cervids are considered to be definitive hosts of giant liver fluke:

- white-tailed deer *Odocoileus virginianus*
- wapiti *Cervus elaphus canadensis*
- Rocky Mountain elk *Cervus elaphus nelsoni*
- Roosevelt elk *Cervus elaphus roosevelti*
- caribou *Rangifer tarandus*
- black-tailed deer *Odocoileus hemionus columbianus*
- mule deer *Odocoileus hemionus hemionus*
- red deer *Cervus elaphus elaphus*
- fallow deer *Dama dama*

It is generally known that *F. magna* is of the North American origin where it co-evolved with ancestral *Odocoileus* sp. White-tailed deer has significantly contributed to maintenance and spread of fascioloidosis in North America, and till now it represents one of the most frequent definitive hosts of giant liver fluke. In general, white-tailed deer tolerate *F. magna* infection without significant clinical signs (Pybus 2001).

In North America, *F. magna* was found in naturally infected white-tailed deer coming from all enzootic regions except for NQL (see Table 3.1 and references therein). The most frequent occurrence of fascioloidosis was determined in SAS enzootic region throughout broad spectrum of southeastern US states, with the highest prevalence (64–84 %) in Texas (Foreyt and Todd 1972; Foreyt et al. 1977). White-tailed deer was also attractive "import article", which was introduced to European parks, enclosures and reservations in the 19th and 20th centuries. In Europe, fascioloidosis was detected in white-tailed deer in the Czech Republic (Erhardová-Kotrlá 1971). Besides white-tailed deer, wapiti and caribou have significantly contributed to distribution of *F. magna* within and between enzootic regions in North America. Both cervids acquired giant liver fluke from white-tailed deer in overlapping regions with sympatric occurrence of different cervids (Kennedy et al. 1999; Bazsalovicsová et al. 2015).

Taxonomy of the genus *Cervus* is not univocal and different authors accept different scientific names and terminology. In general, *Cervus elaphus* is supposed to include many subspecies, including the most frequent "Old World" cervid, red deer *Cervus elaphus elaphus*, and common North American species, wapiti *Cervus elaphus canadensis*. In North America, there are numerous subspecies of *C. elaphus*, including Rocky Mountain elk *Cervus elaphus nelsoni* and Roosevelt elk *Cervus elaphus roosevelti* (Bryant and Maser 1982). Some authors consider red deer and wapiti to be separate species, *Cervus elaphus* and *Cervus canadensis*, respectively (e.g. Groves 2006). In order to avoid any misunderstanding in terminology of deer, we use the original scientific names of all cervids as provided in the reference literature.

Wapiti (Fig. 3.1), one of the largest species of family Cervidae, was found to be infected with *F. magna* mainly in foothills and mountain endemic areas of the Rocky Mountain trench (RMT) enzootic region, in Canadian provinces Montana and Alberta (Banff National Park) (see Table 3.1 and references therein). In NPC enzootic region, fascioloidosis was detected in wapiti from British Columbia and Oregon (Table 3.1). The prevalence detected in RMT and NPC enzootic regions reached up to 80–100 % (Whiting and Tessaro 1994; Hood et al. 1997; Pybus et al. 2015). Sporadic occurrence of *F. magna* in wapiti was detected in Cuba, where it was imported from North America (Lorenzo et al. 1989).

Others "New World" cervids susceptible to *F. magna* infection are Rocky Mountain elk, Roosevelt elk, caribou, black-tailed deer and mule deer (see Table 3.1 and references therein). While Rocky Mountain elk and mule deer were found mainly in RMT region, Roosevelt elk and black-tailed deer infected with *F. magna* were detected in coastal states and provinces of NPC region.

Along with wapiti and white-tailed deer, third dominant definitive host of *F. magna* is reindeer or caribou, terrestrial herbivore of many northern ecosystems (Pollock et al. 2009). The George River herd is the largest caribou population in eastern Canada (mainly Labrador), and represents the only endemic caribou herd in North America infected with *F. magna* (Wobeser et al. 1985; Pollock et al. 2009). Caribou infected with *F. magna* was found only in NQL region, where the dynamics of fascioloidosis is similar to dynamics in populations of wapiti and white-tailed deer (Lankester and Luttich 1988). Excessive number of flukes may lead to mortality even in this type of definitive host (Pybus 2001).

In Europe, the most frequent and dominat definitive host of *F. magna* is red deer. Similarly to white-tailed deer and wapiti in North America, red deer plays an important role in maintaining and spread of fascioloidosis in Europe. Red deer infected with *F. magna* was found in all European natural foci (IT, CZ-PL and DFF) in all countries (see Table 3.1 and references therein). A very high prevalence, reaching up to 100 % was determined in Italy (Balbo et al. 1987), Czech Republic (Erhardová-Kotrlá 1971) and Danube floodplain forests (Rajský et al. 2002; Ursprung and Prosl 2011). Fallow deer is second ruminant species known as

Table 3.1 Spectrum of naturally infected final hosts (Definitive hosts; all family Cervidae) with *F. magna*

Definitive host[a]	Natural habitat	Continent	NA region[b] EU focus[c]	US state, CA province[d] EU country[e]	P (%)	References
White-tailed deer *Odocoileus virginianus*	North and South America	North America	NPC	British Columbia	28	Pybus et al. (2015)
		North America	RMT	Alberta	n.i.	Swales (1935)
		North America	RMT	Alberta	2	Pybus (1990)
		North America	RMT	Alberta	44	Pybus et al. (2015)
		North America	RMT	Montana	n.i.	Aiton (1938) c.i. Pybus (2001)
		North America	GLR	Minnesota	n.i.	Fenstermacher et al. (1943)
		North America	GLR	Minnesota	n.i.	Bazsalovicsová et al. (2015)
		North America	GLR	New York	n.i.	Stiles and Hassall (1894) c.i. Pybus (2001)
		North America	SAS	Florida	n.i.	Dinaburg (1939) c.i. Pybus (2001)
		North America	SAS	Florida	n.i.	Bazsalovicsová et al. (2015)
		North America	SAS	Georgia	n.i.	Bazsalovicsová et al. (2015)
		North America	SAS	Kentucky	n.i.	Lydeard et al. (1989)
		North America	SAS	Louisiana	n.i.	Bazsalovicsová et al. (2015)
		North America	SAS	Mississippi	n.i.	Bazsalovicsová et al. (2015)
		North America	SAS	North Carolina	73	Flowers (1996)
		North America	SAS	South Carolina	n.i.	Dinaburg (1939) c.i. Pybus (2001)
		North America	SAS	South Carolina	30	Lydeard et al. (1989)
		North America	SAS	South Carolina	25.6	Lydeard et al. (1989)
		North America	SAS	South Carolina	11.73	Steele (2008)
		North America	SAS	South Carolina	n.i.	Bazsalovicsová et al. (2015)
		North America	SAS	Tennessee	41.9	Lydeard et al. (1989)
		North America	SAS	Tennessee	53.3	Lydeard et al. (1989)
		North America	SAS	Texas	n.i.	Olsen (1949)

(continued)

Table 3.1 (continued)

Definitive host[a]	Natural habitat	Continent	NA region[b] EU focus[c]	US state, CA province[d] EU country[e]	P (%)	References
		North America	SAS	Texas	69.7	Foreyt and Todd (1972)
		North America	SAS	Texas	64–84	Foreyt et al. (1977)
		North America	SAS	13 Southeastern states	12.8	Pursglove et al. (1977)
		Europe	CZ-PL	Czech Republic	n.i.	Erhardová-Kotrlá (1971)
Wapiti *Cervus elaphus canadensis*	North America, Eastern Asia	North America	NPC	Oregon	n.i.	Dutson et al. (1967)
		North America	NPC	British Columbia	n.i.	Flook and Stenton (1969)
		North America	NPC	British Columbia	77–100	Pybus et al. (2015)
		North America	RMT	Alberta	n.i.	Swales (1935)
		North America	RMT	Alberta	n.i.	Flook and Stenton (1969)
		North America	RMT	Alberta	50	Kingscote et al. (1987)
		North America	RMT	Alberta	80	Whiting and Tessaro (1994)
		North America	RMT	Alberta	3.2–33.3	Kennedy et al. (1999)
		North America	RMT	Alberta	53–79	Pybus et al. (2015)
		North America	RMT	Alberta	n.i.	Bazsalovicsová et al. (2015)
		North America	RMT	Montana	n.i.	Butler (1938) c.i. Pybus (2001)
		North America	RMT	Montana	4–100	Hood et al. (1997)
		North America	RMT	Saskatchewan	n.i.	Wobeser et al. (1985)
		Latin America	n.r.	Cuba	n.i.	Lorenzo et al. (1989)
Rocky Mountain elk *Cervus elaphus nelsoni*	Western North America, Rocky Mountains	North America	RMT	Alberta	29	Pybus (1990)
		North America	RMT	Alberta	93	Pybus et al. (1991)
		Europe	IT	Italy	n.i.	Bassi (1875) c.i. Pybus (2001)
Roosevelt elk *Cervus elaphus roosevelti*	Western North America, Alaska	North America	NPC	British Columbia	n.i.	Bazsalovicsová et al. (2015)
		North America	NPC	Washington	n.i.	Schwartz and Mitchell (1945)

(continued)

Table 3.1 (continued)

Definitive host[a]	Natural habitat	Continent	NA region[b] EU focus[c]	US state, CA province[d] EU country[e]	P (%)	References
Caribou, reindeer *Rangifer tarandus*	Northern Europe, North America, Siberia	North America	NQL	Quebec	n.i.	Choquette et al. (1971)
		North America	NQL	Labrador	58	Lankester and Luttich (1988)
		North America	NQL	Labrador	78	Pollock et al. (2009)
		North America	NQL	Labrador	n.i.	Bazsalovicsová et al. (2015)
Black-tailed deer *Odocoileus hemionus columbianus*	Western North America, Alaska	North America	NPC	British Columbia	n.i.	Hadwen (1916) c.i. Pybus (2001)
		North America	NPC	British Columbia	n.i.	Cowan (1946)
		North America	NPC	Oregon	n.i.	Bazsalovicsová et al. (2015)
Mule deer *Odocoileus hemionus hemionus*	Western North America	North America	NPC	British Columbia	4	Pybus et al. (2015)
		North America	RMT	Alberta	14	Pybus (1990)
		North America	RMT	Alberta	6	Pybus et al. (2015)
		North America	RMT	Montana	n.i.	Senger (1963)
Red deer *Cervus elaphus elaphus*	Europe, Western and Central Asia	Europe	IT	Italy	n.i.	Bassi (1875) c.i. Pybus (2001)
		Europe	IT	Italy	51.8	Lanfranchi et al. (1984/85)
		Europe	IT	Italy	50–100	Balbo et al. (1987)
		Europe	IT	Italy	n.i.	Balbo et al. (1989)
		Europe	IT	Italy	n.i.	Králová-Hromadová et al. (2011)
		Europe	CZ-PL	Czech Republic	81–100	Erhardová-Kotrlá (1971)
		Europe	CZ-PL	Czech Republic	n.i.	Kolář (1978)
		Europe	CZ-PL	Czech Republic	4–95	Novobilský et al. (2007)
		Europe	CZ-PL	Czech Republic	n.i.	Králová-Hromadová et al. (2011)
		Europe	CZ-PL	Poland	n.i.	Ślusarski (1955)
		Europe	CZ-PL	Poland	n.i.	Pyziel et al. (2014)
		Europe	CZ-PL	Poland	n.i.	Demiaszkiewicz et al. (2015)
		Europe	CZ-PL	Poland	n.i.	Králová-Hromadová et al. (2015)
		Europe	DFF	Austria	66.7	Winkelmayer and Prosl (2001)

(continued)

Table 3.1 (continued)

Definitive host[a]	Natural habitat	Continent	NA region[b] EU focus[c]	US state, CA province[d] EU country[e]	P (%)	References
		Europe	DFF	Austria	15.8	Ursprung et al. (2006)
		Europe	DFF	Austria	13–100	Ursprung and Prosl (2011)
		Europe	DFF	Hungary	≥90	Majoros and Sztojkov (1994)
		Europe	DFF	Hungary	21.1–65.1	Giczi (2008)
		Europe	DFF	Hungary	n.i.	Králová-Hromadová et al. (2011)
		Europe	DFF	Slovakia	70–80	Rajský et al. (1995)
		Europe	DFF	Slovakia	90	Rajský et al. (1996)
		Europe	DFF	Slovakia	70	Špakulová et al. (1997)
		Europe	DFF	Slovakia	91.3	Rajský et al. (2002)
		Europe	DFF	Slovakia	17.39	Rajský et al. (2006)
		Europe	DFF	Slovakia	n.i.	Králová-Hromadová et al. (2011)
		Europe	DFF	Croatia	n.i.	Marinculić et al. (2002)
		Europe	DFF	Croatia	54.1	Janicki et al. (2005)
		Europe	DFF	Croatia	20–80	Slavica et al. (2006)
		Europe	DFF	Croatia	53.3	Rajković-Janje et al. (2008)
		Europe	DFF	Croatia	4.05	Rajković-Janje et al. (2008)
		Europe	DFF	Croatia	0–48	Florijančić et al. (2010)
		Europe	DFF	Croatia	n.i.	Králová-Hromadová et al. (2011)
		Europe	DFF	Croatia	57.4	Severin et al. (2012)
		Europe	n.d.	Germany	n.i.	Salomon (1932)
		Europe	n.d.	Germany	70	Rehbein et al. (2012)
		Europe	n.d.	Germany	4.9	Plötz et al. (2015)

(continued)

Table 3.1 (continued)

Definitive host[a]	Natural habitat	Continent	NA region[b] EU focus[c]	US state, CA province[d] EU country[e]	P (%)	References
Fallow deer *Dama dama*	Western Eurasia	Europe	IT	Italy	n.i	Bassi (1875) c.i. Pybus (2001)
		Europe	CZ-PL	Czech Republic	n.i.	Ullrich (1930)
		Europe	CZ-PL	Czech Republic	21.6–31.9	Erhardová-Kotrlá (1971)
		Europe	CZ-PL	Czech Republic	15.8	Chroust and Chroustová (2004)
		Europe	CZ-PL	Czech Republic	4–95	Novobilský et al. (2007)
		Europe	CZ-PL	Poland	n.i.	Demiaszkiewicz et al. (2015)
		Europe	n.d.	Poland	n.i.	Karamon et al. (2015)
		Europe	DFF	Austria	n.i.	Pfeiffer (1983)
		Europe	DFF	Serbia	52.2	Marinković et al. (2013)
		Europe	n.d.	Germany	10.2	Plötz et al. (2015)

NA North America, *CA* Canada, *EU* Europe, *P (%)* prevalence of *Fascioloides magna* infection, *NPC* northern Pacific coast, *RMT* Rocky Mountain trench, *GLR* Great Lakes region, *NQL* northern Quebec and Labrador, *SAS* Gulf coast, lower Mississippi and southern Atlantic seaboard, *IT* Italy, *CZ-PL* Czech Republic and southwestern Poland, *DFF* Danube floodplain forests, *FM Fascioloides magna*, *n.i.* not indicated in the respective literature, *n.r.* not relevant, *n. d.* not detected, *c.i.* cited in

[a]Order of definitive hosts follows the one indicated by Pybus (2001)

[b]North American enzootic regions are listed in west-east and north-south directions

[c]European natural foci are listed chronologically, according to first discoveries in the respective focus

[d]US states and Canadian provinces are listed alphabetically within the respective enzootic region

[e]European countries are listed chronologically, according to first discoveries in the respective country

Fig. 3.1 Wapiti (*Cervus elaphus canadensis*) from Banff National Park, Alberta, Canada. (*Photo* I. Králová-Hromadová)

definitive host of *F. magna* in Europe; fascioloidosis in fallow deer was detected in all natural foci (Table 3.1); the highest prevalence was reported from Czech Republic reaching up to 95 % (Novobilský et al. 2007).

The prevalence of fascioloidosis in definitive hosts is age-dependent. While young hosts are rarely infected, infection is increasing in older age classes (Flook and Stenton 1969; Foreyt et al. 1977; Lankester and Luttich 1988; Mulvey and Aho 1993). Prevalence of adult flukes is comparable in both sexes in wapiti (Pybus 2001), white-tailed deer (Foreyt et al. 1977), and caribou (Lankester and Luttich 1988).

3.1.2 Dead-End Hosts

In dead-end hosts, giant liver fluke can reach the liver but the parasite matures very rarely and few produced eggs are usually not released into the bile system, intestine and further to the external environment. Contrary to definitive hosts, dead-end hosts do not contribute to maintenance of the infection and spread of propagative stages of *F. magna*. Fascioloidosis in this type of hosts may have a lethal effect. Dead-end hosts represent taxonomically diverse category, in particular:

Family Cervidae

- moose *Alces alces*
- sika deer *Cervus nippon*
- sambar deer *Cervus unicolor*

Family Bovidae

- cattle *Bos taurus*
- bison *Bison bison*
- yak *Bos grunniens*
- blue bull *Boselaphus tragocamelus*
- muscox *Ovibus moschatus*

Family Equidae

- horse *Equus* sp.

Family Suidae

- wild boar *Sus scrofa*
- domestic swine *Sus scrofa* f. *domestica*

Family Tayassuidae

- collared peccary *Pecari tajacu*

Family Camelidae

- llama *Lama glama*

Moose represents one of the most frequent dead-end hosts of the family Cervidae. Fascioloidosis in this type of host was detected in NPC, RMT and GLR enzootic regions (see Table 3.2 and references therein), with the highest prevalence determined in British Columbia (63 %; Pybus et al. 2015) and Minnesota (89 %; Murray et al. 2006). In Europe, sika deer was found to be infected with *F. magna* in Czech Republic and Germany (Erhardová-Kotrlá 1971; Rehbein et al. 2012; Plötz et al. 2015).

Regarding domestic ruminants, a cattle represents the most common dead-end host of *F. magna* in North America and Europe. Giant liver fluke mature in the liver, eggs are produced, but stay trapped within hepatic parenchyma and do not enter bile ducts (Foreyt and Todd 1976). Fascioloidosis in cattle causes chronic liver lesions (Price 1953), but the infection is usually not lethal (Lankester 1974; Foreyt and Todd 1976; Foreyt and Parish 1990). Cattle infected with *F. magna* was reported in all North American enzootic regions except for NQL, and in two European countries, Italy and Czech Republic (see Table 3.2 and references therein). A sporadic occurrence was detected in South Africa (Boomker and Dale-Kuys 1977) and Australia (Arundel and Hamir 1982). Within the family Bovidae, *F. magna* infections were also found in bison and yak from Alberta (RMT) (Cameron 1923 c.i. Pybus 2001; Swales 1935), in muskox from Quebec (NQL) (Bazsalovicsová et al. 2015), and in blue bull from Italy (Bassi 1875 c.i. Pybus 2001).

A rather rare dead-end host of *F. magna* is wild boar; infections were documented in Italy (Balbo et al. 1987, 1989) and Texas, with high prevalence ranging from 51.7 % (Foreyt and Todd 1972) to 69 % (Foreyt et al. 1975). Dangerous for pigs may be feeding on pastures contaminated by eggs of infected white-tailed deer, or other definitive hosts (Schwartz et al. 1993). Wild boar does not shed *F. magna*

Table 3.2 Spectrum of naturally infected final hosts (Dead-end hosts) with *F. magna*

Dead-end host[a]	Family	Natural habitat	Continent	NA region[b] EU focus[c]	US state CA province[d] EU country[e]	P (%)	References
Moose *Alces alces*	Cervidae	Canada, Alaska, Scandinavia, Rusia	North America	NPC	British Columbia	n.i.	Hilton (1930) c.i. Pybus (2001)
			North America	NPC	British Columbia	n.i.	Cowan (1951)
			North America	NPC	British Columbia	63	Pybus et al. (2015)
			North America	RMT	Alberta	4	Pybus (1990)
			North America	RMT	Alberta	52	Pybus et al. (2015)
			North America	RMT	Saskatchewan	n.i.	Wobeser et al. (1985)
			North America	GLR	Manitoba	n.i.	Lankester (1974)
			North America	GLR	Minnesota	n.i.	Fenstermacher (1934) c.i. Pybus (2001)
			North America	GLR	Minnesota	89	Murray et al. (2006)
			North America	GLR	Minnesota	17.4; 5.2	Peterson et al. (2013)
			North America	GLR	North Dakota	0–19.6	Maskey (2011)
			North America	GLR	Ontario	n.i.	Kingscote (1950)
Sika deer *Cervus nippon*	Cervidae	Eastern Asia	Europe	CZ-PL	Czech Republic	4	Erhardová-Kotrlá (1971)
			Europe	n.d.	Germany	37.5	Rehbein et al. (2012)
			Europe	n.d.	Germany	0	Plötz et al. (2015)
Sambar deer *Cervus unicolor*	Cervidae	Southern and Southeastern Asia	Europe	IT	Italy	n.i.	Bassi (1875) c.i. Pybus (2001)

(continued)

Table 3.2 (continued)

Dead-end host[a]	Family	Natural habitat	Continent	NA region[b] EU focus[c]	US state CA province[d] EU country[e]	P (%)	References
Cattle *Bos taurus*	Bovidae	Worldwide distribution, domesticated	North America	NPC	British Columbia	n.i.	Hilton (1930) c.i. Pybus (2001)
			North America	RMT	Alberta	n.i.	Swales (1935)
			North America	RMT	Montana	17.24	Knapp et al. (1992)
			North America	RMT	Saskatchewan	n.i.	Wobeser and Schumann (2014)
			North America	GLR	Michigan	0.41–13.9	Schillhorn van Veen (1987)
			North America	SAS	Texas	n.i.	Francis (1891) c.i. Pybus (2001)
			North America	SAS	Texas	38.3	Foreyt and Todd (1972)
			Europe	IT	Italy	3.4	Lanfranchi et al. (1984/85)
			Europe	IT	Italy	3.7	Balbo et al. (1987)
			Europe	CZ-PL	Czech Republic	n.i.	Záhoř et al. (1966)
			Europe	CZ-PL	Czech Republic	9.1–21.1	Chroustová et al. (1980)
			Europe	CZ-PL	Czech Republic	n.i.	Leontovyč et al. (2014)
			South Africa	n.r.	n.i.	n.i.	Boomker and Dale-Kuys (1977)
			Australia	n.r.	n.i.	n.i.	Arundel and Hamir (1982)
Bison *Bison bison*	Bovidae	Western Europe, Central Asia, North America	North America	RMT	Alberta	n.i.	Cameron (1923) c.i. Pybus (2001)
			North America	RMT	Alberta	n.i.	Swales (1935)
Yak *Bos grunniens*	Bovidae	Southern Asia	North America	RMT	Alberta	n.i.	Swales (1935)
Blue bull *Boselaphus tragocamelus*	Bovidae	Southern Asia	Europe	IT	Italy	n.i.	Bassi (1875) c.i. Pybus (2001)

(continued)

Table 3.2 (continued)

Dead-end host[a]	Family	Natural habitat	Continent	NA region[b] EU focus[c]	US state CA province[d] EU country[e]	P (%)	References
Muskox *Ovibus moschatus*	Bovidae	Canadian Arctic, Greenland	North America	NQL	Quebec	n.i.	Bazsalovicsová et al. (2015)
Horse *Equus* sp.	Equidae	Worldwide distribution, domesticated	North America	GLR	Minnesota	n.i.	McClanahan et al. (2005)
			Europe	IT	Italy	5.7	Balbo et al. (1987)
Wild boar *Sus scrofa*	Suidae	Eurasia, North Africa, Greater Sunda Islands, USA	North America	SAS	Texas	51.7	Foreyt and Todd (1972)
			North America	SAS	Texas	69	Foreyt et al. (1975)
			North America	SAS	Texas	n.i.	Schwartz et al. (1993)
			Europe	IT	Italy	n.i.	Balbo et al. (1987)
			Europe	IT	Italy	n.i.	Balbo et al. (1989)
Collared peccary *Pecari tajacu*	Tayassuidae	North, Central, South America	North America	SAS	Texas	1	Samuel and Low (1970)
Llama *Lama glama*	Camelidae	South America, domesticated	North America	GLR	Minnesota	n.i.	Conboy et al. (1988)

NA North America, *CA* Canada, *EU* Europe, *P (%)* prevalence of *Fascioloides magna* infection, *NPC* northern Pacific coast, *RMT* Rocky Mountain trench, *GLR* Great Lakes region, *NQL* northern Quebec and Labrador, *SAS* Gulf coast, lower Mississippi and southern Atlantic seaboard, *IT* Italy, *CZ-PL* Czech Republic and southwestern Poland, *DFF* Danube floodplain forests, *n.i.* not indicated in the respective literature, *n.r.* not relevant, *n.d.* not detected, *c.i.* cited in

[a]Order of dead-end hosts follows the one indicated by Pybus (2001)

[b]North American enzootic regions are listed in west-east and north-south directions

[c]European natural foci are listed chronologically, according to first discoveries in the respective focus

[d]US states and Canadian provinces are listed alphabetically within the respective enzootic region

[e]European countries are listed chronologically, according to first discoveries in the respective country

eggs to environment and, therefore, does not contribute to further spread of fascioloidosis (Foreyt et al. 1975). Rare *F. magna* infections were detected in horse (McClanahan et al. 2005) and llama (Conboy et al. 1988) from Minnesota (GLR), in horse from Italy (Balbo et al. 1987), and in collared peccary from Texas (SAS) (Samuel and Low 1970).

3.1.3 Aberrant Hosts

In aberrant hosts, giant liver fluke can not successfully complete migration within the ruminant host; parasite may move up to the liver but formation of pseudocysts is not successful. These hosts may often die due to tissue damage, which is associated with migration of immature flukes through peritoneal, thoracic or abdominal cavities. According to Pybus (2001), aberrant hosts are mainly domestic, but also free living ruminants:

Family Bovidae

- domestic sheep *Ovis aries*
- domestic goat *Capra hircus*
- chamois *Rupicapra rupicapra*
- bighorn sheep *Ovis canadensis*
- mouflon *Ovis orientalis*

Family Cervidae

- roe deer *Capreolus capreolus*

In aberrant hosts, such as sheep and goat, unrestricted migration of fluke through the liver, lungs and peritoneal cavities is typical. It results to massive tissue damage, usually with fatal effects caused even by relatively low intensity of infection (Conboy and Stromberg 1991). Fascioloidosis in domestic sheep and goat was documented both in North America and Europe (see Table 3.3 and references therein). Roe deer, the only cervid species recognized as an aberrant host, was found to be infected with *F. magna* only in Europe, in particular CZ-PL and DFF (see Table 3.3 and references therein). The highest prevalence of fascioloidosis in roe deer was detected in Czech Republic (70–80 %; Erhardová-Kotrlá 1971) and Slovakia (60 %; Rajský et al. 2002).

3.2 Experimentally Infected Final Hosts

The experimental infections of different types of final hosts with *F. magna* were aimed to determine the clinical signs, pathological changes and immunological responses of final hosts under controlled experimental conditions. The major

Table 3.3 Spectrum of naturally infected final hosts (ABERRANT HOSTS) with *F. magna*

Aberrant host[a]	Family	Natural habitat	Continent	NA region[b] EU focus[c]	US state, CA province[d] EU country[e]	P (%)	References
Domestic sheep *Ovis aries*	Bovidae	Worldwide distribution, domesticated	North America	NPC	Oregon	34.3	Foreyt and Hunter (1980)
			North America	RMT	Montana	n.i.	Hall (1914) c.i. Pybus (2001)
			North America	GLR	Wisconsin	n.i.	Campbell and Todd (1954)
			North America	SAS	Texas	n.i.	Olsen (1949)
			Europe	IT	Italy	n.i.	Bassi (1875) c.i. Pybus (2001)
Domestic goat *Capra hircus*	Bovidae	Worldwide distribution, domesticated	North America	SAS	Texas	n.i.	Olsen (1949)
			Europe	IT	Italy	n.i.	Bassi (1875) c.i. Pybus (2001)
Roe deer *Capreolus capreolus*	Cervidae	Eurasia	Europe	CZ-PL	Czech Republic	n.i.	Záhoř (1965)
			Europe	CZ-PL	Czech Republic	70–80	Erhardová-Kotrlá (1971)
			Europe	CZ-PL	Czech Republic	9.1	Chroust and Chroustová (2004)
			Europe	CZ-PL	Poland	n.i.	Demiaszkiewicz et al. (2015)
			Europe	DFF	Austria	n.i.	Winkelmayer and Prosl (2001)
			Europe	DFF	Austria	n.i.	Ursprung et al. (2006)
			Europe	DFF	Hungary	3.7	Giczi (2008)
			Europe	DFF	Slovakia	60	Rajský et al. (2002)
			Europe	DFF	Slovakia	n.i.	Rajský et al. (2006)
			Europe	n.d.	Germany	20	Rehbein et al. (2012)

NA North America, *CA* Canada, *EU* Europe, *P (%)* prevalence of *Fascioloides magna* infection, *NPC* northern Pacific coast, *RMT* Rocky Mountain trench, *GLR* Great Lakes region, *NQL* northern Quebec and Labrador, *SAS* Gulf coast, lower Mississippi and southern Atlantic seaboard, *IT* Italy, *CZ-PL* Czech Republic and southwestern Poland, *DFF* Danube floodplain forests, *n.i.* not indicated in the respective literature, *n.d.* not detected, *c.i.* cited in

[a]Order of aberrant hosts follows the one indicated by Pybus (2001)

[b]North American enzootic regions are listed in west-east and north-south directions

[c]European natural foci are listed chronologically, according to first discoveries in the respective focus

[d]US states and Canadian provinces are listed alphabetically within the respective enzootic region

[e]European countries are listed chronologically, according to first discoveries in the respective country

Table 3.4 Results on experimental infections of different types of final hosts infected with *F. magna*

Experimental animal	No. of meta	Primary localization	Localization in other organs	FM eggs in faeces	Conclusions of experiments	References
White-tailed deer (DH)	500	Liver	Lungs	n.i.	Early prepatent FM infection revealed mild transitory anemia and extensive migration of immature flukes	Presidente et al. (1980)
	50–500	Liver	n.i.	Yes	No significant clinical signs	Foreyt and Todd (1979)
	40–500	Liver	n.i.	Yes	Confirmed definitive host	Foreyt and Todd (1976)
	100	n.i.	n.i.	Yes	Infected animal remained clinically healthy	Foreyt (1996a)
	500	Liver	Lungs Abdominal cavity Thoracic cavity	Yes	No significant clinical signs	Foreyt (1992)
Wapiti (DH)	2,000	Liver	Peritoneal cavity	n.i.	Lethal effect	Foreyt (1996a)
	250	Liver	n.i.	Yes	No significant clinical signs	Foreyt (1996a)
Mule deer (DH)	500	Liver	Lungs Pleural cavity Peritoneal cavity	No	Poor appetite, depression, weakness, lethal effect	Foreyt (1992)
	50	Liver	n.i.	Yes	Confirmed definitive host	Foreyt (1996b)
	250	Liver	n.i.	Yes	Depression, droopy ears, weakness, lethal effect	Foreyt (1996b)

(continued)

Table 3.4 (continued)

Experimental animal	No. of meta	Primary localization	Localization in other organs	FM eggs in faeces	Conclusions of experiments	References
Fallow deer (DH)	32–120	Liver	Peritoneal cavity Abdominal cavity	Yes	Poor appetite, apathy, paroxysm, lethal effect	Erhardová-Kotrlá and Blažek (1970)
Moose (DEH)	50–225	Liver	n.i.	n.i.	Depressed appetite 10 days after infection, later no significant clinical signs	Lankester and Foreyt (2011)
Cattle (DEH)	60	Liver	n.i.	No	No significant clinical signs, later meteorism, poor appetite	Erhardová-Kotrlá and Blažek (1970)
	10–500	Liver	Lungs	No	Eggs retained in the liver	Foreyt and Todd (1976)
	1,000	Liver	Abdominal cavity	n.i.	Well tolerated infection, no clinical signs observed	Conboy and Stromberg (1991)
Bison (DEH)	600	None	n.i.	No	Not developed FM infection	Foreyt and Drew (2010)
Llama (DEH)	250	Liver	n.i.	No	Clinical signs similar to those detected in cattle	Foreyt and Parish (1990)
Sheep (AH)	8–200	Liver	Lungs Abdominal cavity	No	Lethal effect	Foreyt and Todd (1976)
Bighorn sheep (AH)	50	Liver	n.i.	No	Lethal effect	Erhardová-Kotrlá and Blažek (1970)
	50–100	Liver	Lungs Peritoneal cavity	n.i.	Lethal effect	Foreyt (1996a)

(continued)

Table 3.4 (continued)

Experimental animal	No. of meta	Primary localization	Localization in other organs	FM eggs in faeces	Conclusions of experiments	References
Chamois (AH)	250–320	Liver	Lungs	No	No significant clinical signs, lethal effect	Erhardová-Kotrlá and Blažek (1970)
Guinea pig (AH)	20	Liver	Lungs Abdominal cavity Thoracic cavity Sceletal muscle Subcutaneous tissue	n.i.	Lethal effect Infection similar to FM infection in sheep Guinea pigs—possible models for FM infection studies in sheep	Conboy and Stromberg (1991)
	10	Liver	Peritoneal cavity	n.i.	Lethal effect	Foreyt and Todd (1979)

DH definitive host, *DEH* dead-end host, *AH* aberrant host, *No. of meta* number of metacercariae used for infection, *FM Fascioloides magna*, n.i. not indicated contribution in this field has to be addressed to William J. Foreyt from Washington State University in Pullman, Washington, USA.

Experimental infections were studied in all types of final hosts of *F. magna*, in particular in definitive hosts (white-tailed deer, wapiti, mule deer and fallow deer), dead-end hosts (moose, cattle, bison and llama), and aberrant hosts (sheep, bighorn sheep, chamois and guinea pig) (see Table 3.4 and references therein). The animals were infected with dose of 8–2,000 infective stages (metacercariae) per animal, most frequently in the number of 200–500. The main monitored parameters were localization of parasite, ability of parasite to reach the maturity, detection of *F. magna* eggs in host's faeces and determination of overall clinical signs of infected hosts (including lethal effect). In some cases, haematological and blood chemistry values were determined, as well. The results on experimental infections can be correlated with data known from natural infections; the classification of final hosts can thus be determined in more details.

The majority of experiments were performed in white-tailed deer, the primary definitive host of *F. magna*. The fluke was localized in liver, with occasional occurence in lungs, abdominal and thoracic cavities (see Table 3.4 and references therein). The important finding was detection of *F. magna* eggs in feaces of white-tailed deer, what clearly demonstrates its ability to provide suitable conditions for parasite's maturity, production of eggs, their release into the external environment and consequent spread of the infection.

White-tailed deers were without significant clinical signs and were confirmed to be definitive host for *F. magna*. Presidente et al. (1980) studied haematological values of white-tails infected experimentally with *F. magna*. A reduction of erythrocytes and an elevation of reticulocytes, macrocytic cells and eosinophils were detected in the mentioned study. On the other hand, serum proteins, albumins and globulins remained under the physiological values. Another study confirmed decrease of haemoglobin, increase of total serum proteins, β- and γ-globulin fractions (Foreyt and Todd 1979).

In wapiti, localization of parasite in liver and peritoneal cavity was detected (Foreyt 1996a). Using 250 metacercariae as an infectious dose, no significant clinical signs were recorded and eggs of *F. magna* were detected in faeces. However, a massive infection of wapiti (2,000 metacercariae as infectious dose) was proved to have a lethal effect (Foreyt 1996a). Lethal effect was determined also after experimental infection of mule deer (Foreyt 1992, 1996b), in which giant liver fluke was primarily localized in liver, but even in lungs, pleural and peritoneal cavities; eggs of *F. magna* were found in faecal samples of mule deer. At the end of experimental infection of fallow deer, poor appetite, apathy and paroxysm appeared, and increased γ-globulins and hypoalbuminaemia were detected. Infection had a lethal effect; flukes were found in the liver, peritoneal and abdominal cavities (Erhardová-Kotrlá and Blažek 1970).

In dead-end hosts (moose, cattle and llama) experimentally infected with *F. magna*, dominant localization of the parasite was liver (Erhardová-Kotrlá and Blažek 1970; Foreyt and Todd 1976; Foreyt and Parish 1990; Conboy and Stromberg 1991; Lankester and Foreyt 2011), although presence of *F. magna* was confirmed also in lungs and abdominal cavity of cattle (Foreyt and Todd 1976; Conboy and Stromberg 1991). The important finding in all studied dead-end hosts was that eggs were not released into faeces, but were found to be retained in liver (Foreyt and Todd 1976). These results correspond to definition of dead-end hosts, which do not contribute to spread of propagative stages of the parasite into external environment. No clinical signs were detected in moose (Lankester and Foreyt 2011) and cattle (Conboy and Stromberg 1991). Fascioloidosis did not develop in experimentally infected bison (Foreyt and Drew 2010).

Experimental infections in aberrant hosts (sheep and bighorn sheep) revealed presence of parasite in liver, but also in lungs, peritoneal and abdominal cavities; fascioloidosis in this type of hosts had a lethal effect (Erhardová-Kotrlá and Blažek 1970; Foreyt and Todd 1976; Foreyt 1996a). In chamois experimentally infected with *F. magna*, no clinical signs were observed during the whole period of experiment. However, on the 138th day after infestation the chamois suddenly died and flukes were found in liver and lungs (Erhardová-Kotrlá and Blažek 1970). As generally known for aberrant hosts, eggs were not detected in faeces.

Already small dose of infective metacercariae (10 and 20) resulted in lethal effect of fascioloidosis in guinea pig, in which natural infections were not determined. As expected, the infection was quite extensive; except for liver, flukes were determined in lungs, peritoneal, abdominal and thoracic cavities, and even in skeletal muscles and subcutaneous tissues (Foreyt and Todd 1979; Conboy and Stromberg 1991). The response observed in guinea pigs was similar to that reported in sheep, suggesting the suitability of the guinea pig as a model for *F. magna* infection in sheep (Conboy and Stromberg 1991).

References

Aiton JF (1938) Enlarged spleen in white-tailed deer at Glacier National Park. Transactions of the North American Wildlife Conference 3:890–892. Cited in Pybus MJ (2001) Liver flukes. In: Samuel WM, Pybus MJ, Kocan AA (eds) Parasitic diseases of wild mammals, 2nd edn. Iowa State University Press, Ames

Arundel JH, Hamir AN (1982) *Fascioloides magna* in cattle. Aust Vet J 58:35–36

Balbo T, Lanfranchi P, Rossi L, Meneguz PG (1987) Health management of a red deer population infected by *Fascioloides magna* (Bassi, 1875) Ward, 1917. Ann Fac Med Vet Torino 32:1–13

Balbo T, Rossi P, Meneguz PG (1989) Integrated control of *Fascioloides magna* infection in Northern Italy. Parassitologia 31:137–144

Bassi R (1875) Sulla cachessia ittero-verminosa, o marciaia, causta dei Cervi, causata dal *Distomum magnum*. Il Medico Veterinario 4:497–515. Cited in Pybus MJ (2001) Liver flukes. In: Samuel WM, Pybus MJ, Kocan AA (eds) Parasitic diseases of wild mammals, 2nd edn. Iowa State University Press, Ames

Bazsalovicsová E, Králová-Hromadová I, Štefka J, Minárik G, Bokorová S, Pybus M (2015) Genetic interrelationships of North American populations of giant liver fluke *Fascioloides magna*. Parasit Vectors 8:1–15. doi:10.1186/s13071-015-0895-1

Boomker J, Dale-Kuys JC (1977) First report of *Fascioloides magna* (Bassi, 1875) in South Africa. Onderstepoort J Vet Res 44:49–52

Bryan LD, Maser C (1982) Classification and distribution. In: Thomas JW, Toweill DE, Metz DP (eds) Elk of North America: ecology and management. Stackpole Books, Harrisburg, Pennsylvania

Butler WJ (1938) Wild animal disease investigation. Montana Livestock Sanitary Board 1:18–19. Cited in Pybus MJ (2001) Liver flukes. In: Samuel WM, Pybus MJ, Kocan AA (eds) Parasitic diseases of wild mammals, 2nd edn. Iowa State University Press, Ames

Cameron AE (1923) Notes on buffalo: Anatomy, pathological conditions, and parasites. Brit Vet J 79:331–336. Cited in Pybus MJ (2001) Liver flukes. In: Samuel WM, Pybus MJ, Kocan AA (eds) Parasitic diseases of wild mammals, 2nd edn. Iowa State University Press, Ames

Campbell WC, Todd AC (1954) Natural infections of *Fascioloides magna* in Wisconsin sheep. J Parasitol 40:100

Choquette LPE, Gibson GG, Simard B (1971) *Fascioloides magna* (Bassi, 1875) Ward, 1917 (Trematoda) in woodland caribou, *Rangifer tarandus caribou* (Gmelin), of northeastern Quebec, and its distribution in wild ungulates in Canada. Can J Zool 49:280–281

Chroust K, Chroustová E (2004) Motolice obrovská (*Fascioloides magna*) u spárkaté zvěře v jihočeských lokalitách. Veterinářství 54:296–304 (in Czech)

Chroustová E, Hůlka J, Jaroš J (1980) Prevence a terapie fascioloidózy skotu bithionolsulfoxidem. Vet Med (Praha) 25:557–563 (in Czech)

Conboy GA, Stromberg BE (1991) Hematology and clinical pathology of experimental *Fascioloides magna* infection in cattle and guinea pigs. Vet Parasitol 40:241–255

Conboy GA, O'Brien TD, Stevens DL (1988) A natural infection of *Fascioloides magna* in a llama (*Lama glama*). J Parasitol 74:345–346

Cowan IM (1946) Parasites, diseases, injuries, and anomalies of the Columbian black-tailed deer, *Odocoileus hemionus columbianus* (Richardson), in British Columbia. Can J Res 24:71–103

Cowan IM (1951) The diseases and parasites of big game mammals of western Canada. Proc Ann Game Convention 5:37–64

Demiaszkiewicz AW, Kuligowska I, Pyziel AM, Lachowicz J, Kowalczyk R (2015) Extension of occurrence area of the American fluke *Fascioloides magna* in south-western Poland. Ann Parasitol 61:93–96

Dinaburg AG (1939) Helminth parasites collected from deer, *Odocoileus virginianus* in Florida. Proc Helminthol Soc Wash 6:102–104. Cited in Pybus MJ (2001) Liver flukes. In: Samuel WM, Pybus MJ, Kocan AA (eds) Parasitic diseases of wild mammals, 2nd edn. Iowa State University Press, Ames

Dutson VJ, Shaw JN, Knapp SE (1967) Epizootiologic factors of *Fascioloides magna* (Trematoda) in Oregon and southern Washington. Am J Vet Res 28:853–860

Erhardová-Kotrlá B (1971) The occurrence of *Fascioloides magna* (Bassi, 1875) in Czechoslovakia. Czechoslovak Academy of Sciences, Prague

Erhardová-Kotrlá B, Blažek K (1970) Artificial infestation caused by the fluke *Fascioloides magna*. Acta Vet Brno 39:287–295

Fenstermacher R (1934) Diseases affecting moose. Alumni Q 22:81–94. Cited in Pybus MJ (2001) Liver flukes. In: Samuel WM, Pybus MJ, Kocan AA (eds) Parasitic diseases of wild mammals, 2nd edn. Iowa State University Press, Ames

Fenstermacher R, Olsen WO, Pomeroy BS (1943) Some diseases of white-tailed deer in Minnesota. Cornell Vet 33:323–332

Flook DR, Stenton JE (1969) Incidence and abundance of certain parasites in wapiti in the national parks of the Canadian Rockies. Can J Zool 47:795–803. doi:10.1139/z69-138

Florijančić T, Ozimec S, Opačak A, Bošković I, Jelkić D, Marinculić A, Janicki Z (2010) Importance of the Danube River in spreading the infection of red deer with *Fascioloides magna* in eastern Croatia. Paper presented at 38th IAD Conference, Dresden, Germany, 22–25 June 2010

Flowers J (1996) Notes on the life history of *Fascioloides magna* (Trematoda) in North Carolina. J Elisha Mitch Sci S 112:115–118

Foreyt WJ (1992) Experimental *Fascioloides magna* infections of mule deer (*Odocoileus hemionus hemionus*). J Wildl Dis 28:183–187. doi:10.7589/0090-3558-28.2.183

Foreyt WJ (1996a) Susceptibility of bighorn sheep (*Ovis canadensis*) to experimentally-induced *Fascioloides magna* infections. J Wildlife Dis 32:556–559

Foreyt WJ (1996b) Mule deer (*Odocoileus hemionus*) and elk (*Cervus elaphus*) as experimental definitive hosts for *Fascioloides magna*. J Wildlife Dis 32:603–606

Foreyt WJ, Todd AC (1972) The occurrence of *Fascioloides magna* and *Fasciola hepatica* together in the livers of naturally infected cattle in South Texas, and the incidence of the flukes in cattle, white-tailed deer, and feral hogs. J Parasitol 58:1010–1011

Foreyt WJ, Todd AC (1976) Liver flukes in cattle: prevalence, distribution and experimental treatment. Vet Med Small Anim Clin 71:816–822

Foreyt WJ, Todd AC (1979) Selected clinicopathologic changes associated with experimentally induced *Fascioloides magna* infection in white-tailed deer. J Wildl Dis 15:83–89

Foreyt WJ, Hunter RL (1980) Clinical *Fascioloides magna* infection in sheep in Oregon on pasture shared by Columbian white-tailed deer. Am J Vet Res 41:1531–1532

Foreyt WJ, Parish S (1990) Experimental infection of liver flukes (*Fascioloides magna*) in a llama (*Lama glama*). J Zoo Wildl Med 21:468–470

Foreyt WJ, Drew ML (2010) Experimental infection of liver flukes, *Fasciola hepatica* and *Fascioloides magna*, in bison (*Bison bison*). J Wildlife Dis 46:283–286. doi:10.7589/0090-3558-46.1.283

Foreyt WJ, Todd AC, Foreyt K (1975) *Fascioloides magna* (Bassi, 1875) in feral swine from southern Texas. J Wildlife Dis 11:554–559. doi:10.7589/0090-3558-11.4.554

Foreyt WJ, Samuel WM, Todd AC (1977) *Fascioloides magna* in white-tailed deer (*Odocoileus virginianus*): observation of the pairing tendency. J Parasitol 63:1050–1052. doi:10.2307/3279843

Francis M (1891) Liver flukes. Tex AES Bull 18:123–136. Cited in Pybus MJ (2001) Liver flukes. In: Samuel WM, Pybus MJ, Kocan AA (eds) Parasitic diseases of wild mammals, 2nd edn. Iowa State University Press, Ames

Giczi E (2008) *Fascioloides magna* (Bassi, 1875) infection of Hungarian red deer and roe deer stock and the possibility of protection. Dissertation, University of West Hungary

Groves C (2006) The genus *Cervus* in eastern Eurasia. Eur J Wildl Res 52:14–22. doi:10.1007/s10344-005-0011-5

Hadwen S (1916) A new host for *Fasciola magna*, Bassi, together with observation on the distribution of *Fasciola hepatica*, L. in Canada. J Am Vet Med Assoc 49:511–515. Cited in Pybus MJ (2001) Liver flukes. In: Samuel WM, Pybus MJ, Kocan AA (eds) Parasitic diseases of wild mammals, 2nd edn. Iowa State University Press, Ames

Hall MC (1914) Society proceedings of the Helminthological Society of Washington. J Parasitol 1:106. Cited in Pybus MJ (2001) Liver flukes. In: Samuel WM, Pybus MJ, Kocan AA (eds) Parasitic diseases of wild mammals, 2nd edn. Iowa State University Press, Ames

Hilton G (1930) Report of the Veterinary Director General, Department of Agriculture, Ottawa, Canada. Cited in Pybus MJ (2001) Liver flukes. In: Samuel WM, Pybus MJ, Kocan AA (eds) Parasitic diseases of wild mammals, 2nd edn. Iowa State University Press, Ames

Hood BR, Rognlie MC, Knapp SE (1997) Fascioloidiasis in game-ranched elk from Montana. J Wildl Dis 33:882–885. doi:10.7589/0090-3558-33.4.882

Janicki Z, Konjević D, Severin K (2005) Monitoring and treatment of *Fascioloides magna* in semi-farm red deer husbandry in Croatia. Vet Res Commun 29:83–88. doi:10.1007/s11259-005-0027-z

Karamon J, Larska M, Jasik A, Sell B (2015) First report of the giant liver fluke (*Fascioloides magna*) infection in farmed fallow deer (*Dama dama*) in Poland—pathomorphological changes and molecular identification. Bull Vet Inst Pulawy 59:339–344. doi:10.1515/bvip-2015-0050

Kennedy MJ, Acorn RC, Moraiko DT (1999) Survey of *Fascioloides magna* in farmed wapiti in Alberta. Can Vet J 40:252–254

Kingscote AA (1950) Liver rot (Fascioloidiasis) in ruminants. Can J Comp Med Vet Sci 14:203–208

Kingscote BF, Yates WDG, Tiffin GB (1987) Diseases of wapiti utilizing cattle range in southwestern Alberta. J Wildl Dis 23:86–91

Knapp SE, Dunkel AM, Han K, Zimmerman LA (1992) Epizootiology of fascioliasis in Montana. Vet Parasitol 42:241–246

Kolář Z (1978) Příspěvek k léčbě fascioloidózy u jelení zvěře. Veterinářství 28:276–277 (in Czech)

Králová-Hromadová I, Bazsalovicsová E, Štefka J, Špakulová M, Vávrová S, Szemes T, Tkach V, Trudgett A, Pybus M (2011) Multiple origins of European populations of the giant liver fluke *Fascioloides magna* (Trematoda: Fasciolidae), a liver parasite of ruminants. Int J Parasitol 41:373–383. doi:10.1016/j.ijpara.2010.10.010

Králová-Hromadová I, Bazsalovicsová E, Demiaszkiewicz A (2015) Molecular characterization of *Fascioloides magna* (Trematoda: Fascioloidae) from south-western Poland based on mitochondrial markers. Acta Parasitol 60:544–547. doi:10.1515/ap-2015-0077

Lanfranchi P, Tolari F, Forletta R, Meneguz PG, Rossi L (1984/85) The red deer as reservoir of parasitic and infectious pathogens for cattle. Ann Fac Med Vet Torino 30:1–17

Lankester MW (1974) *Parelaphostrongylus tenuis* (Nematoda) and *Fascioloides magna* (Trematoda) in moose of southeastern Manitoba. Can J Zool 52:235–239

Lankester MW, Luttich S (1988) *Fascioloides magna* (Trematoda) in woodland caribou (*Rangifer tarandus caribou*) of the George River herd, Labrador. Can J Zool 66:475–479. doi:10.1139/z88-067

Lankester MW, Foreyt WJ (2011) Moose experimentally infected with giant liver fluke (*Fascioloides magna*). Alces 47:9–15

Leontovyč R, Košťáková M, Siegelová V, Melounová K, Pankrác J, Vrbová K, Horák P, Kašný M (2014) Highland cattle and *Radix labiata*, the hosts of *Fascioloides magna*. BMC Vet Res 10:1–7. doi:10.1186/1746-6148-10-41

Lorenzo M, Ramirez P, Mendez M, Alonso M, Ramos R (1989) Reporte de *Fascioloides magna*, Bassi, 1875, parasitando un wápiti (*Cervus canadensis*) en Cuba. Revista Cubana de Ciencias Veterinarias 20:263–266

Lydeard C, Mulvey M, Aho JM, Kennedy PK (1989) Genetic variability among natural populations of the liver fluke *Fascioloides magna* in white-tailed deer, *Odocoileus virginianus*. Can J Zool 67:2021–2025. doi:10.1139/z89-287

Majoros G, Sztojkov V (1994) Appearance of the large American liver fluke *Fascioloides magna* (Bassi, 1875) (Trematoda: Fasciolata) in Hungary. Parasit Hung 27:27–38

Marinculić A, Džakula N, Janicki Z, Hardy Z, Lučinger S, Živičnjak T (2002) Appearance of American liver fluke (*Fascioloides magna*, Bassi, 1875) in Croatia—a case report. Vet Arhiv 72:319–325

Marinković D, Kukolj V, Aleksić-Kovačević S, Jovanović M, Knežević M (2013) The role of hepatic myofibroblasts in liver cirrhosis in fallow deer (*Dama dama*) naturally infected with giant liver fluke (*Fascioloides magna*). BMC Vet Res 9:45. doi:10.1186/1746-6148-9-45

Maskey JJ (2011) Giant liver fluke in North Dakota moose. Alces 47:1–7

McClanahan SL, Stromberg BE, Hayden DW, Averbeck GA, Wilson JH (2005) Natural infection of a horse with *Fascioloides magna*. J Vet Diagn Invest 17:382–385. doi:10.1177/104063870501700415

Mulvey M, Aho JM (1993) Parasitism and mate competition: liver flukes in white-tailed deer. Oikos 66:187–192. doi:10.2307/3544804

Murray DL, Cox EW, Ballard WB, Whitlaw HA, Lenarz MS, Custer TW, Barnett T, Fuller TK (2006) Pathogens, nutritional deficiency, and climate influences on a declining moose population. Wildlife Monogr 166:1–30. doi:10.2193/0084-0173(2006)166

Novobilský A, Horáčková E, Hirtová L, Modrý D, Koudela B (2007) The giant liver fluke *Fascioloides magna* (Bassi 1875) in cervids in the Czech Republic and potential of its spreading to Germany. Parasitol Res 100:549–553. doi:10.1007/s00436-006-0299-4

Olsen OW (1949) White-tailed deer as a reservoir of the large American liver fluke. Vet Med 44:26–30

Peterson WJ, Lankester MW, Kie JG, Bowyer RT (2013) Geospatial analysis of giant liver flukes among moose: effects of white-tailed deer. Acta Theriol 58:359–365. doi:10.1007/s13364-013-0130-4

Pfeiffer H (1983) *Fascioloides magna*: Erster Fund in Österreich. Wien Tierarztl Monat 70:168–170 (in German)

Plötz C, Rehbein S, Bamler H, Reindl H, Pfister K, Scheuerle MC (2015) *Fascioloides magna*—epizootiology in a deer farm in Germany. Berl Munch Tierarztl Wochenschr 128:177–182. doi:10.2376/0005-9366-128-177

Pollock B, Penashue B, McBurney S, Vanleeuwen J, Daoust PY, Burgess NM, Tasker AR (2009) Liver parasites and body condition in relation to environmental contaminants in caribou (*Rangifer tarandus*) from Labrador, Canada. Arctic 62:1–12

Presidente PJA, McCraw BM, Lumsden JH (1980) Pathogenicity of immature *Fascioloides magna* in white-tailed deer. Can J Comparat Med 44:423–432

Price EW (1953) The fluke situation in American ruminants. J Parasitol 39:119–134

Pursglove SR, Prestwood AK, Ridgeway TR, Hayes FA (1977) *Fascioloides magna* infection in white-tailed deer of southeastern United States. J Am Vet Med Assoc 171:936–938

Pybus MJ (1990) Survey of hepatic and pulmonary helminths of wild cervids in Alberta, Canada. J Wildl Dis 26:453–459. doi:10.7589/0090-3558-26.4.453

Pybus MJ (2001) Liver flukes. In: Samuel WM, Pybus MJ, Kocan AA (eds) Parasitic diseases of wild mammals, 2nd edn. Iowa State University Press, Ames

Pybus MJ, Onderka DK, Cool N (1991) Efficacy of triclabendazole against natural infections of *Fascioloides magna* in wapiti. J Wildlife Dis 27:599–605

Pybus MJ, Butterworth EW, Woods JG (2015) An expanding population of the giant liver fluke (*Fascioloides magna*) in elk (*Cervus canadensis*) and other ungulates in Canada. J Wildl Dis 51:431–445. doi:10.7589/2014-09-235

Pyziel AM, Demiaszkiewicz AW, Kuligowska I (2014) Molecular identification of *Fascioloides magna* (Bassi, 1875) from red deer from south-western Poland (Lower Silesian Wilderness) on the basis of internal transcribed spacer 2 (ITS-2). Pol J Vet Sci 17:523–525. doi:10.2478/pjvs-2014-0077

Rajković-Janje R, Bosnić S, Rimac D, Gojmerac T (2008) The prevalence of American liver fluke *Fascioloides magna* (Bassi 1875) in red deer from Croatian hunting grounds. Eur J Wildl Res 54:525–528. doi:10.1007/s10344-007-0163-6

Rajský D, Patus A, Bielik J (1995) Záchyt cicavice obrovskej (*Fascioloides magna* Bassi, 1875) v rámci monitoringu bioty v oblasti Vodného diela Gabčíkovo. In: Kontrišová O, Kočík K, Bublinec E (eds) Zborník referátov, Monitorovanie a hodnotenie stavu životného prostredia. Technická univerzita, Zvolen (in Slovak)

Rajský D, Patus A, Špakulová M (1996) Rozšírenie cicavice obrovskej (*Fascioloides magna* Bassi, 1875) v jelenej chovateľskej oblasti J–I Podunajská. In: Zborník referátov a príspevkov medzinárodnej konferencie 1996. Výskumný ústav živočíšnej výroby, Nitra (in Slovak)

Rajský D, Čorba J, Várady M, Špakulová M, Cabadaj R (2002) Control of fascioloidosis (*Fascioloides magna* Bassi, 1875) in red deer and roe deer. Helminthologia 39:67–70

Rajský D, Dubinský P, Krupicer I, Sabo R, Sokol J (2006) Výskyt propagačných štádií *Fascioloides magna* a iných helmintov vo fekáliách jelenej zveri z okresov hraničiacich s riekami Dunaj a Morava. Slov Vet Čas 31:177–180 (in Slovak)

Rehbein S, Hamel D, Reindl H, Visser M, Pfister K (2012) *Fascioloides magna* and *Ashworthius sidemi*—two new parasites in wild ungulates in Germany. In: Program and abstracts of the XI European multicolloquium of parasitology (EMOP XI), Cluj-Napoca, Romania, 25–29 July 2012

Salomon S (1932) *Fascioloides magna* bei deutschem Rotwild. Berl Tierärztl Wochenschr 48:627–628 (in German)

Samuel WM, Low WA (1970) Parasites of the collared peccary from Texas. J Wildl Dis 6:16–23

Schillhorn van Veen TW (1987) Prevalence of *Fascioloides magna* in cattle and deer in Michigan. J Am Vet Med Assoc 191:547–548

Schwartz JE, Mitchell GE (1945) The Roosevelt Elk on the Olympic Peninsula, Washington. J Wildlife Manage 9:295–319. doi:10.2307/3796372

Schwartz WL, Lawhorn DB, Montgomery E (1993) *Fascioloides magna* in a feral pig. Swine Health Prod 1:27

Senger CM (1963) Some parasites of Montana deer. Montana Wildl Autumn:5–13

Severin K, Mašek T, Janicki Z, Konjević D, Slavica A, Marinculić A, Martinković F, Vengušt G, Džaja P (2012) Liver enzymes and blood metabolites in a population of free-ranging red deer (*Cervus elaphus*) naturally infected with *Fascioloides magna*. J Helminthol 86:190–196. doi:10.1017/S0022149X1100023X

Slavica A, Florijančić T, Janicki Z, Konjević D, Severin K, Marinculić A, Pintur K (2006) Treatment of fascioloidosis (*Fascioloides magna*, Bassi 1875) in free ranging and captive red deer (*Cervus elaphus* L.) at eastern Croatia. Vet Arhiv 76:9–18

Ślusarski W (1955) Studia nad europejskimi przedstawicielami przywry *Fasciola magna* (Bassi, 1875) Stiles, 1894. Acta Parasitol Pol 3:1–59 (in Polish)

Špakulová M, Čorba J, Varády M, Rajský D (1997) Bionomy, distribution and importance of giant liver fluke (*Fascioloides magna*), an important parasite of free-living ruminants. Vet Med 42:139–148

Steele E (2008) Prevalence of the large liver fluke, *Fascioloides magna*, in the white-tailed deer in South Carolina. In: Paper presented at the 4th annual USC upstate research symposium, University of South Carolina, Spartanburg, 11 April 2008

Stiles CW, Hassall A (1894) The anatomy of the large American fluke (*Fasciola magna*), and a comparison with other species of the genus *Fasciola*. J Comp Med Vet Arch 15:161–178, 225–243, 299–313, 407–417, 457–462. Cited in Pybus MJ (2001) Liver flukes. In: Samuel WM, Pybus MJ, Kocan AA (eds) Parasitic diseases of wild mammals, 2nd edn. Iowa State University Press, Ames

Swales WE (1935) The life cycle of *Fascioloides magna* (Bassi, 1875), the large liver fluke of ruminants, in Canada. Can J Res 12:177–215. doi:10.1139/cjr35-015

Ullrich K (1930) Über das Vorkommen von seltenen oder wenig bekannten Parasiten der Säugetiere und Vögel in Böhmen und Mähren. Prag Arch Tiermed 10:19–43 (in German)

Ursprung J, Prosl H (2011) Vorkommen und Bekämpfung des Amerikanischen Riesenleberegels (*Fascioloides magna*) in den österreichischen Donauauen östlich von Wien 2000–2010. Wien Tierarztl Monat 98:275–284 (in German)

Ursprung J, Joachim A, Prosl H (2006) Epidemiology and control of the giant liver fluke, *Fascioloides magna*, in a population of wild ungulates in the Danubian wetlands east of Vienna. Berl Munch Tierarztl Wochenschr 119:316–323 (in German)

Whiting TL, Tessaro SV (1994) An abattoir study of tuberculosis in a herd of farmed elk. Can Vet J 35:497–501

Winkelmayer R, Prosl H (2001) Riesenleberegel – jetzt auch bei uns? Österreichisches Weidwerk 3:42–44 (in German)

Wobeser BK, Schumann F (2014) *Fascioloides magna* infection causing fatal pulmonary hemorrhage in a steer. Can Vet J 55:1093–1095

Wobeser G, Gajadhar AA, Hunt HM (1985) *Fascioloides magna*: Occurrence in Saskatchewan and distribution in Canada. Can Vet J 26:241–244

Záhoř Z (1965) Výskyt velké motolice (*Fascioloides magna* Bassi, 1875) u srnčí zvěře. Veterinářství 15:329–324 (in Czech)

Záhoř Z, Prokeš C, Vítovec L (1966) Nález vajíček motolice *Fascioloides magna* (Bassi, 1875) a fascioloidózních změn v játrech skotu. Vet Med (Praha) 39:397–404 (in Czech)

Chapter 4
Intermediate Hosts of *Fascioloides magna*

Abstract The complexity of the life cycle of *Fascioloides magna* and its ability to invade new region is ensured by the presence of suitable intermediate hosts, in particular aquatic pulmonate mollusks, in which larval development of the parasite takes place. This Chapter summarizes intermediate snail hosts of giant liver fluke specific in North America and Europe. In North America, six species of the family Lymnaeidae were found to be naturally infected with *F. magna* (*Lymnaea caperata*, *Lymnaea modicella*, *Stagnicola palustris nuttalliana*, *Pseudosuccinea columella*, *Galba bulimoides techella* and *Fossaria parva*). In Europe, *Galba* (syn. *Lymnaea*) *truncatula*, *Radix labiata* and *Radix peregra* were found to be naturally infected. Besides natural infections, number of snail species were experimentally infected with *F. magna* in order to determine their potential to serve as the intermediate hosts of giant liver fluke. The mature cercariae able to develop into infective metacercariae stages, were detected in snails of the genera *Lymnaea* and *Pseudosuccinea* (family Lymnaeidae) in North America and in lymnaeid genera *Galba*, *Lymnaea*, *Omphiscola*, *Pseudosuccinea* and *Stagnicola* in Europe. It is evident, that broader spectrum of aquatic mollusks is susceptible to *F. magna* infection and may serve as its potential intermediate hosts.

Keywords Giant liver fluke · Intermediate hosts · Freshwater snail · Lymnaeidae · *Lymnaea* · *Galba* · *Radix* · Experimental infection · Natural infection

4.1 General Characterization of Intermediate Snail Hosts

The complex life cycle of *F. magna* requires suitable intermediate hosts, pulmonate freshwater gastropod mollusks, in which larval development of the parasite takes place (see Sect. 1.3; Fig. 1.2). In North America and Europe, specific spectrum of snail species of the family Lymnaeidae was detected to be naturally infected with *F. magna* (see Tables 4.1, 4.2 and references therein).

I. Králová-Hromadová et al., *The Giant Liver Fluke, Fascioloides magna: Past, Present and Future Research*, SpringerBriefs in Animal Sciences,
DOI 10.1007/978-3-319-29508-4_4

Table 4.1 Spectrum of naturally infected intermediate snail hosts (Lymnaeidae) with *F. magna* in North America

Snail species	CA province US state	Enzootic region	References
Lymnaea caperata	USA/Minnesota	GLR	Griffiths (1959)
	USA/Minnesota	GLR	Laursen and Stromberg (1993)
	USA/Montana	RMT	Knapp et al. (1992)
Lymnaea modicella	USA/n.i.	n.i.	Krull (1933, 1934) c.i. Swales (1935)
	USA/Minnesota	GLR	Laursen and Stromberg (1993)
Stagnicola palustris nuttalliana	USA/Montana	RMT	Swales (1935)
Pseudosuccinea columella	USA/Montana	RMT	Krull (1933, 1934) c.i. Swales (1935)
Galba bulimoides techella	USA/Texas	SAS	Sinitsin (1930) c.i. Swales (1935)
Fossaria parva	CA/Alberta	RMT	Swales (1935)

CA Canada, *US* United States, *GLR* Great Lakes region, *RMT* Rocky Mountain trench, *SAS* Gulf coast, lower Mississippi, and southern Atlantic seaboard, *n.i.* not indicated in the respective literature, *c.i.* cited in

Table 4.2 Spectrum of naturally infected intermediate snail hosts (Lymnaeidae) with *F. magna* in Europe

Snail species	Country	Natural focus	References
Galba (syn. *Lymnaea*) *truncatula*	Czech Republic	CZ-PL	Erhardová (1961)
	Czech Republic	CZ-PL	Erhardová-Kotrlá (1971)
	Czech Republic	CZ-PL	Chroust and Chroustová (2004)
	Czech Republic	CZ-PL	Faltýnková et al. (2006)
	Czech Republic	CZ-PL	Kašný et al. (2012)
	Czech Republic	CZ-PL	Leontovyč et al. (2014)
	Austria	DFF	Hörweg et al. (2011)
	Austria	DFF	Haider et al. (2012)
	Slovakia	DFF	Rajský et al. (1996)
	Hungary	DFF	Majoros and Sztojkov (1994)
Radix labiata	Czech Republic	CZ-PL	Leontovyč et al. (2014)
Radix peregra	Czech Republic	CZ-PL	Faltýnková et al. (2006)

CZ-PL Czech Republic and southwestern Poland, *DFF* Danube floodplain forests

Taxonomic classification and systematics of the family Lymnaeidae, and snails in general, is rather complicated and underwent several revisions. As generally accepted in current modern taxonomy and systematics, the most effective strategy for an accurate classification of the species is traditional alfa-taxonomy based on morphological descriptions with the combination of DNA-based methods. The morphology of reproductive system and shell morphometry are crucial markers for identification of snails; however, they are not generally applicable markers for all taxa (e.g. shell morphology is not suitable for delimitation of species of the genus *Radix*; Pfenninger et al. 2006). Therefore, molecular tools, in particular internal transcribed spacer 2 of the ribosomal DNA (ITS2 rDNA), were applied in the family Lymnaeidae as effective marker of molecular taxonomy (Mas-Coma et al. 2009; Huňová et al. 2012; Leontovyč et al. 2014).

Recently, many snail species have underwent taxonomic revisions (e.g. *Stagnicola palustris* and *Omphiscola glabra* were transferred to the genus *Lymnaea*; Correa et al. 2010; Novobilský et al. 2012); different scientific names of some snail species were simultaneously used by several authors. We summarized the data on naturally and experimentally infected intermediate snail hosts of *F. magna* indicating the original scientific name of the snail, as provided in the reference literature.

Apart from natural infections, several attempts to infect different mollusks under experimental conditions were performed by researchers on both continents. Sporocysts, mother rediae and daughter rediae represent non-infectious larval intramolluscan stages. On the other hand, mature cercariae released from snails are able to encyst in exogenous environment on aquatic vegetation, and may develop to metacercariae, the stage infective for final host. Therefore, full development of *F. magna* intramolluscan stages and production of mature cercariae were the main criteria for an assessment of the potential of snail species to serve as an intermediate host.

The successful infection of intermediate host by *F. magna* depends mainly on the susceptibility of selected snail, the infectivity of miracidia (Smyth and Halton 1983), and favourable environmental, ecological and physical factors (e.g. temperature, humidity etc.) (Rapsch et al. 2008). The optimal temperature range for infections is 15–30 °C, when a sufficiently moist environment assists the development of the first larval stage, miracidium. Changes in temperature and moisture may also considerably influence the complete embryonation process during summer field conditions (Pybus 2001). Besides suitable hydrological conditions may determine the population density of intermediate hosts (Rajský et al. 2002).

One of the key factors influencing susceptibility of snail is epidermal mucus covering the surface of snail host. It may serve as an important barrier for an attempt of *F. magna* miracidium to penetrate into snails. *Fascioloides magna*-incompatible snails possess a potent cytotoxic protein-like factor in the mucus, which is absent in *F. magna*-compatible snails. This factor could play a significant role in mediating larval trematode-snail compatibility (Coyne et al. 2015).

4.2 Natural Infections

North America Different larval stages of giant liver fluke were found in six snail species of the family Lymnaeidae, in particular *Lymnaea caperata*, *L. modicella*, *Stagnicola palustris nuttalliana*, *Pseudosuccinea columella*, *Galba bulimoides techella* and *Fossaria parva* in three enzootic regions (GLR, RMT and SAS) (see Table 4.1 and references therein). In natural conditions, co-existence of different snail species susceptible to *F. magna* can significantly increase the risk of infection and its further spread to final hosts. For example in Alberta (RMT), the Canadian province with one of the highest prevalence of fascioloidosis in free-living ruminants, four species of lymnaeid snails are known to be suitable intermediate hosts for *F. magna* either in natural or in experimental conditions (Kennedy et al. 1999).

Europe Comparing with North America, lower number of intermediate snail hosts has been detected in Europe (Table 4.2). The most frequent and the only confirmed intermediate host is air-breathing freshwater snail *Galba* (syn. *Lymnaea) truncatula* (Fig. 4.1), in which the larval development of the parasite can be completed. *Galba truncatula* was proved to produce mature cercariae and can significantly contribute to successful transmission of infective stages to final hosts (Erhardová-Kotrlá 1971). The snail requires a moderate climate and moisture for its survival and reproduction. *Galba truncatula* is dominant intermediate host of *F. magna* in Europe, what is probably closely related to the adaptation of this snail species to a wide range of ecological conditions and biotops (running waters, backwater systems, river banks, ponds, marsh areas, flooded meadows etc.) (Hörweg et al. 2011; Haider et al. 2012).

Natural infections of *G. truncatula* with *F. magna* were mainly detected in the Czech Republic (Erhardová 1961; Erhardová-Kotrlá 1971; Chroust and Chroustová 2004; Faltýnková et al. 2006; Kašný et al. 2012; Leontovyč et al. 2014), also reported in Danube floodplain forests, including Slovakia (Rajský et al. 1996), Hungary (Majoros and Sztojkov 1994) and Austria (Hörweg et al. 2011; Haider et al. 2012). The mean prevalence of fascioloidosis in *G. truncatula* varied from 0.03–0.23 % in Austria (Hörweg et al. 2011; Haider et al. 2012) up to 30–60 % in Czech Republic (Faltýnková et al. 2006; Kašný et al. 2012; Leontovyč et al. 2014). It is interesting that in Austria, almost 8-fold increase of infection rate in

Fig. 4.1 *Galba* (syn. *Lymnaea) truncatula* from Danube floodplain forests, Slovakia (*Photo* M. Špakulová)

G. truncatula was observed despite an ongoing triclabendazole treatment programme of final hosts (Haider et al. 2012).

Galba truncatula was considered to be the only intermediate host of *F. magna* in Europe for a long time (Erhardová-Kotrlá 1971), until Faltýnková et al. (2006) described naturally infected *R. peregra* in Czech Republic (prevalence 0.08 %). Despite the fact, that *R. peregra* is dominant snail species over *G. truncatula* in Czech Republic, *R. peregra* can produce only immature cercariae of *F. magna* (Erhardová-Kotrlá 1971; Faltýnková et al. 2006).

Radix labiata is another species, in which natural infection of *F. magna* was detected (Leontovyč et al. 2014). Similarly to *R. peregra*, also *R. labiata* produced cercariae unable to encyst. However, *R. labiata* might represent a potential intermediate host of *F. magna* in localities ecologically unsuitable for *G. truncatula* (e.g. localities with acid soils) (Leontovyč et al. 2014). The determination of natural infections of *F. magna* in other snail hosts in Europe indicates that the parasite undergoes a process of adaptation to other mollusks (Faltýnková et al. 2006). There is a threat that a broader spectrum of competent intermediate hosts will be detected in the future.

4.3 Experimental Infections

Experimental *F. magna* infections of different snail species were primarily focused on determination of (i) susceptibility of snails to be infected with *F. magna* miracidia; (ii) development of larval stages (sporocysts and rediae) of the parasite within the infected snail hosts; (iii) the potential of snails to produce mature cercariae able to develop into metacercariae (Foreyt and Todd 1978; Sanabria et al. 2013).

North America A spectrum of experimentally infected snails of families Lymnaeidae and Planorbidae is summarized in Table 4.3. Out of 12 tested species, cercariae and infective stages (metacercariae) were detected in seven lymnaeids; *Lymnaea bulimoides*, *Lymnaea caperata*, *Lymnaea humilis*, *Lymnaea palustris*, *Lymnaea stagnalis*, *Lymnaea umbrosa* and *Pseudosuccinea columella* (see Table 4.3 and references therein). In North America, namely *L. caperata* and *P. columella* represent intermediate hosts with previously determined natural *F. magna* infection (Krull 1933, 1934 c.i. Swales 1935; Griffiths 1959; Knapp et al. 1992; Laursen and Stromberg 1993), so they ability to produce infective stages in experimental conditions was anticipated. The remaining five mollusks represent very probably intermediate hosts with a potential to be infected also naturally. Production of rediae was determined but further development into cercariae and metacercariae was not detected in *Ferrissia fragilis*, the only experimentally infected snail of the family Planorbidae (Flowers 1996). The development and production of infective stages in four remaining lymnaeids (*Fossaria modicella rustica*, *Lymnaea ferruginea*, *Lymnaea modicella* and *Stagnicola palustris*) were

Table 4.3 Spectrum of experimentally infected intermediate snail hosts (Lymnaeidae and Planorbidae*) with *F. magna* in North America

Snail species	Detected stage of parasite	Infective stage	US state	Enzootic region	References
Lymnaea bulimoides	Metacercariae	yes	Wisconsin	GLR	Foreyt and Todd (1978)
Lymnaea caperata	Metacercariae	yes	Wisconsin	GLR	Foreyt and Todd (1978)
	n.i.	n.i.	Montana	RMT	Dunkel et al. (1996)
Lymnaea humilis	Metacercariae	yes	Wisconsin	GLR	Foreyt and Todd (1978)
Lymnaea palustris	Metacercariae	yes	Wisconsin	GLR	Foreyt and Todd (1978)
Lymnaea stagnalis	Metacercariae	yes	Wisconsin	GLR	Foreyt and Todd (1978)
	Cercariae	n.i.	Minnesota	GLR	Wu and Kingscote (1954)
	Cercariae	n.i.	Minnesota	GLR	Griffiths (1973)
	Rediae	n.i.	Minnesota	GLR	Friedl (1961)
Lymnaea umbrosa	Metacercariae	yes	Wisconsin	GLR	Foreyt and Todd (1978)
Pseudosuccinea columella	Metacercariae	yes	North Carolina	SAS	Flowers (1996)
	n.i.	n.i.	USA/n.i.	n.i.	Krull (1933) c.i. Swales (1935)
*Ferrissia fragilis**	Rediae	no	North Carolina	SAS	Flowers (1996)
Fossaria modicella rustica	n.i.	n.i.	USA/n.i.	n.i.	Krull (1933) c.i. Swales (1935)
Lymnaea ferruginea	n.i.	n.i.	Oregon	NPC	Dutson et al. (1967)
	n.i.	n.i.	Washington	NPC	Dutson et al. (1967)
Lymnaea modicella	n.i.	n.i.	n.i.	n.i.	Krull (1934) c.i. Swales (1935)
Stagnicola palustris	n.i.	n.i.	Minnesota	GLR	Griffiths (1962)

GLR Great Lakes region, *RMT* Rocky Mountain trench, *SAS* Gulf coast, lower Mississippi, and southern Atlantic seaboard, *NPC* northern Pacific coast, *n.i.* not indicated in the respective literature, *c.i.* cited in either not determined during the experimental infections, or not indicated in the respective literature (see Table 4.3 and references therein).

Europe Adaptation of *F. magna* to different spectrum of intermediate snail hosts in European natural conditions has been one of the crucial factors for successful establishment of natural foci of giant liver fluke outside North America. Experimental infections, focused on detection of susceptibility of various snail species to *F. magna* infection and determination of a spectrum of potential intermediate hosts of *F. magna*, were carried out mainly by the Czech researchers (Erhardová 1961; Erhardová-Kotrlá 1971; Chroustová 1979; Faltýnková et al. 2006; Novobilský et al. 2007, 2012; Huňová et al. 2012), or under international cooperation with Czech parasitologists (Vignoles et al. 2006, 2014; Rondelaud et al. 2006, 2014; Sanabria et al. 2013).

Spectrum of experimentally infected intermediate snail hosts in Europe is summarized in Table 4.4. Despite the fact that development of *F. magna* in snail species other than *Galba* (syn. *Lymnaea*) *truncatula* is much slower (Erhardová 1961) and the prepatent period is longer (Swales 1935; Erhardová-Kotrlá 1971; Rondelaud et al. 2006), infective stages of mature cercariae were developed in several experimentally tested snails. The majority of experimental infections were carried out using *G. truncatula* (Erhardová 1961; Erhardová-Kotrlá 1971; Faltýnková et al. 2006; Vignoles et al. 2006, 2014; Novobilský et al. 2007; Rondelaud et al. 2006, 2014; Sanabria et al. 2013). The experiments confirmed completion of larval development and production of infective stages in this most frequent naturally infected mollusk in Europe. Mature cercariae/metacercariae were detected also after experimental infection of several other species of family Lymnaeidae, in particular *Lymnaea fuscus* (Novobilský et al. 2012), *Lymnaea palustris* (Chroustová 1979), *Omphiscola glabra* (Rondelaud et al. 2006), *Pseudosuccinea columella* (Novobilský et al. 2007), *Stagnicolla palustris* (Chroustová 1979) and *Radix peregra* (Faltýnková et al. 2006).

In some experimental infections, *G. truncatula*, *L. fuscus*, *O. glabra* and *R. peregra* snails originated from France (or Sweden), while *F. magna* eggs were obtained from Czech Republic (see Table 4.4 and references therein). It was assumed, that allopatry/sympatry of snails and *F. magna* miracidia might influence the larval development of the parasite during the experimental infection and modify the intensity of infection and production of cercariae.

Indeed, the larval development of giant liver fluke was more intense, when both, *G. truncatula* and eggs/miracidia of *F. magna* originated from the Czech Republic (Erhardová-Kotrlá 1971). In contrast, *G. truncatula* from France experimentally infected with eggs/miracidia from Czech Republic showed lower or even absent production of cercariae (Rondelaud et al. 2006).

Development of *F. magna* into non-infective stage of sporocysts (mother and daughter rediae, young cercariae) was determined in lymnaeid species of genera *Lymnaea* (*L. peregra ovata*, *L. peregra peregra* and *L. stagnalis*) and *Radix* (*R. lagotis*, *R. labiata*, *R. peregra peregra* and *R. peregra ovata*) (Erhardová 1961; Erhardová-Kotrlá 1971; Faltýnková et al. 2006; Huňová et al. 2012). Since mature cercariae were not produced by these snail species, they would probably not

Table 4.4 Spectrum of experimentally infected intermediate snail hosts (Lymnaeidae, Physidae* and Succinidae$^{\Delta}$) with *F. magna* in Europe

Snail species	Detected stage of parasite	Infective stage	Country	Enzootic region	References
Galba (syn. *Lymnaea*) *truncatula*	Cercariae, metacercariae	yes	Czech Republic	CZ-PL	Erhardová (1961)
	Metacercariae	yes	Czech Republic	CZ-PL	Novobilský et al. (2007)
	Cercariae, metacercariae	yes	France/CZ[a]	n.r.	Vignoles et al. (2006)
	Cercariae, metacercariae	yes	France/CZ[a]	n.r.	Rondelaud et al. (2006)
	Cercariae, metacercariae	yes	France/CZ[a]	n.r.	Vignoles et al. (2014)
	Cercariae, metacercariae	yes	France/CZ[a]	n.r.	Sanabria et al. (2013)
	Rediae, cercariae	n.i.	France/CZ[a]	n.r.	Rondelaud et al. (2014)
	Rediae, cercariae	no	Czech Republic	CZ-PL	Faltýnková et al. (2006)
	Rediae, cercariae	no	Czech Republic	CZ-PL	Erhardová-Kotrlá (1971)
Lymnaea fuscus	Cercariae, metacercariae	yes	France/Sweden/CZ[b]	n.r.	Novobilský et al. (2012)
Lymnaea palustris	Cercariae, metacercariae	yes	Czech Republic	CZ-PL	Chroustová (1979)
	n.i.	n.i.	Czech Republic	CZ-PL	Erhardová-Kotrlá (1971)
Omphiscola glabra	Cercariae, metacercariae	yes	France/CZ[a]	n.r.	Rondelaud et al. (2006)
Pseudosuccinea columella	Metacercariae	yes	Czech Republic	CZ-PL	Novobilský et al. (2007)
Stagnicola palustris	Encysting cercariae, metacercariae	yes	Czech Republic	CZ-PL	Chroustová (1979)
Radix peregra	Mother and daughter rediae, mature cercariae	no	France/CZ[c]	n.r.	Faltýnková et al. (2006)
	n.i.	n.i.	Czech Republic	CZ-PL	Huňová et al. (2012)
Lymnaea peregra ovata	Mother rediae	no	Czech Republic	CZ-PL	Erhardová (1961)
Lymnaea peregra peregra	Mother rediae	no	Czech Republic	CZ-PL	Erhardová (1961)
Lymnaea stagnalis	Sporocysts	no	Czech Republic	CZ-PL	Erhardová-Kotrlá (1971)
	n.i.	no	Czech Republic	CZ-PL	Faltýnková et al. (2006)

(continued)

Table 4.4 (continued)

Snail species	Detected stage of parasite	Infective stage	Country	Enzootic region	References
Radix lagotis	Rediae	no	Czech Republic	CZ-PL	Huňová et al. (2012)
Radix labiata	Rediae, young cercariae	no	Czech Republic	CZ-PL	Huňová et al. (2012)
Radix peregra peregra	Mother rediae	no	Czech Republic	CZ-PL	Erhardová-Kotrlá (1971)
Radix peregra ovata	Sporocysts, mother rediae	no	Czech Republic	CZ-PL	Erhardová-Kotrlá (1971)
*Physa acuta**	n.i.	no	Czech Republic	CZ-PL	Erhardová-Kotrlá (1971)
Succinea oblonga$^{\Delta}$	n.i.	no	Czech Republic	CZ-PL	Erhardová-Kotrlá (1971)
Succinea putris$^{\Delta}$	n.i.	no	Czech Republic	CZ-PL	Erhardová-Kotrlá (1971)

CZ-PL Czech Republic and southwestern Poland, *CZ* Czech Republic, *n.i.* not indicated in the respective literature, *n.r.* not relevant
[a]eggs/miracidia of *F. magna* originated from Czech Republic and snails from France
[b]eggs/miracidia of *F. magna* originated from Czech Republic and snails from France and Sweden,
[c]eggs/miracidia of *F. magna* originated from Czech Republic and snails from France and Czech Republic

contribute to completion of the life cycle of *F. magna* in natural conditions. The findings on experimental infections in *R. labiata* and *R. peregra* (Erhardová-Kotrlá 1971; Huňová et al. 2012) corroborate data on natural infections in these snail species (Faltýnková et al. 2006; Leontovyč et al. 2014).

Besides family Lymnaeidae, experimental infections were carried out also in species of the families Physidae (*Physa acuta*) and Succinidae (*Succinea oblonga* and *S. putris*) (Erhardová-Kotrlá 1971). However, their potential to serve as intermediate hosts of *F. magna* in Europe was excluded, since miracidia of *F. magna* did not enter these snails. In conclusion, the only declared or potential hosts of giant liver fluke in Europe are species of the family Lymnaeidae.

Experimental infections revealed also very interesting phenomena, such as "age-related resistance" and "parasitic gigantism". The first one was observed in *L. fuscus* infection, when only juvenile snails (measuring less than 3 mm; 1–3 weeks of age) were successfully infected with *F. magna* and produced viable cercariae (Novobilský et al. 2012). Success of *F. magna* infection decreased with age of a snail, as documented by increased shell height. Age-related resistence is probably associated with the progressive development of snails' immune system (Novobilský et al. 2012). "Parasitic gigantism" is explained as intensive growth stimulation of snails during redial and cercarial development (Thompson 1997). This general phenomenom was observed also for *F. magna* infections (Vignoles et al. 2006); contrary, reduced snail growth was detected in *L. fuscus* infected with *F. magna* (Novobilský et al. 2012).

4.4 Intermediate Snail Hosts in Other Continents

Apart from North America and Europe, experimental infections were performed also in snails of family Lymnaeidae from South America (Fig. 4.2). In all these experiments, eggs/miracidia of *F. magna* originated from Czech Republic. The relatively high prevalence of infection of two South American snails, *Lymnaea neotropica* (57.4 %; Argentina) and *Lymnaea viatrix* var. *ventricosa* (45.9 %; Uruguay), and successful development of *F. magna* cercariae and metacercariae in these mollusks indicated, that *F. magna* has the high potential to spread to new territories and adapt to local snail species (Sanabria et al. 2013). Similarly, *Lymnaea cubensis* (Guadeloupe) was able to sustain complete larval development of the parasite (prevalence 28 %), including the shedding of cercariae under experimental conditions (Vignoles et al. 2014). All three South American lymnaeids, as well as *Austropeplea (Lymnaea) tomentosa* in Australia (Foreyt and Todd 1974) can be potential new intermediate hosts of *F. magna*, even though the fascioloidosis has not been recorded in these countries (Sanabria et al. 2013).

Natural infections		Experimental infections	
Lymnaea caperata *Lymnaea modicella* *Stagnicola palustris nuttalliana* *Pseudosuccinea columella* *Galba bulimoides techella* *Fossaria parva*	North America	*Lymnaea bulimoides* *Lymnaea caperata* *Lymnaea humilis* *Lymnaea palustris* *Lymnaea stagnalis* *Lymnaea umbrosa* *Pseudosuccinea columella*	*Ferrissia fragilis* *Fossaria modicella rustica* *Lymnaea ferruginea* *Lymnaea modicella* *Stagnicola palustris*
Galba (Lymnaea) truncatula *Radix labiata* *Radix peregra*	Europe 	*Galba (Lymnaea) truncatula* *Lymnaea fuscus* *Lymnaea palustris* *Omphiscola glabra* *Pseudosuccinea columella* *Stagnicola palustris* *Radix peregra* *Lymnaea peregra ovata* *Lymnaea peregra peregra*	*Lymnaea stagnalis* *Radix lagotis* *Radix labiata* *Radix peregra peregra* *Radix peregra ovata* *Physa acuta* *Succinea oblonga* *Succinea putris*
	South America	*Lymnaea neotropica* (Argentina) *Lymnaea viatrix* var. *ventricosa* (Uruguay) *Lymnaea cubensis* (Guadeloupe)	
	Australia	*Austropeplea (Lymnaea) tomentosa* (Australia)	

Fig. 4.2 The worldwide spectrum of naturally and experimentally infected intermediate snail hosts of *F. magna*. Experimental infections: *in blue* snails with developed mature cercariae or infective stages metacercariae; *in green* snails with developed non-infective larval stages (sporocysts, mother and daughter rediae, immature or young cercariae); *in black* production of infective stages either not determined during the experimental infections, or not indicated in the respective literature

References

Chroustová E (1979) Experimental infection of *Lymnaea palustris* snails with *Fascioloides magna*. Vet Parasitol 5:57–64. doi:10.1016/0304-4017(79)90040-2

Chroust K, Chroustová E (2004) Motolice obrovská (*Fascioloides magna*) u spárkaté zvěře v jihočeských lokalitách. Veterinářství 54:296–304 (in Czech)

Coyne K, Laursen JR, Yoshino TP (2015) In vitro effects of mucus from the mantle of compatible (*Lymnaea elodes*) and incompatible (*Helisoma trivolvis*) snail hosts on *Fascioloides magna* miracidia. J Parasitol 101:351–357. doi:10.1645/14-606.1

Correa AC, Escobar JS, Durand P, Renaud F, David P, Jarne P, Pointier JP, Hurtrez-Boussès S (2010) Bridging gaps in the molecular phylogeny of the Lymnaeidae (Gastropoda: Pulmonata), vectors of fascioliasis. BMC Evol Biol 10:381. doi:10.1186/1471-2148-10-381

Dunkel AM, Rognlie MC, Johnson GR, Knapp SE (1996) Distribution of potential intermediate hosts for *Fasciola hepatica* and *Fascioloides magna* in Montana, USA. Vet Parasitol 62:63–70. doi:10.1016/0304-4017(95)00859-4

Dutson VJ, Shaw JN, Knapp SE (1967) Epizootiologic factors of *Fascioloides magna* (Trematoda) in Oregon and southern Washington. Am J Vet Res 28:853–860

Erhardová B (1961) Vývojový cyklus motolice obrovské *Fasciola magna* v podmínkách ČSSR. Zool Listy 10:9–16 (in Czech)

Erhardová-Kotrlá B (1971) The occurrence of *Fascioloides magna* (Bassi, 1875) in Czechoslovakia. Czechoslovak Academy of Sciences, Prague

Faltýnková A, Horáčková E, Hirtová L, Novobilský A, Modrý D, Scholz T (2006) Is *Radix peregra* a new intermediate host of *Fascioloides magna* (Trematoda) in Europe? Field and experimental evidence. Acta Parasitol 51:87–90. doi:10.2478/s11686-006-0013-9

Flowers J (1996) Notes on the life history of *Fascioloides magna* (Trematoda) in North Carolina. J Elisha Mitch Sci S 112:115–118

Foreyt WJ, Todd AC (1974) *Lymnaea tomentosa* from Australia, an experimental intermediate host of the large American liver fluke, *Fascioloides magna*. Aust Vet J 50:471–472

Foreyt WJ, Todd AC (1978) Experimental infection of lymnaeid snails in Wisconsin with miracidia of *Fascioloides magna* and *Fasciola hepatica*. J Parasitol 64:1132–1134. doi:10.2307/3279747

Friedl FE (1961) Studies on larval *Fascioloides magna*. I. Observations on the survival of rediae *in vitro*. J Parasitol 47:71–75. doi:10.2307/3274982

Griffiths HJ (1959) *Stagnicola (Hinckleyia) caperata* (Say), a natural intermediate host for *Fascioloides magna* (Bassi, 1875), in Minnesota. J Parasitol 45:146. doi:10.2307/3286517

Griffiths HJ (1962) Fascioloidiasis of cattle, sheep, and deer in Northern Minnesota. J Am Vet Med Assoc 140:342–347

Griffiths HJ (1973) *Galba modicella* and *Lymnaea stagnalis* as experimental intermediate hosts for *Fascioloides magna* in Minnesota. J Parasitol 59:121

Haider M, Hörweg C, Liesinger K, Sattmann H, Walochnik J (2012) Recovery of *Fascioloides magna* (Digenea) population in spite of treatment programme? Screening of *Galba truncatula* (Gastropoda, Lymnaeidae) from Lower Austria. Vet Parasitol 187:445–451. doi:10.1016/j.vetpar.2012.01.032

Hörweg C, Prosl H, Wille-Piazzai W, Joachim A, Sattmann H (2011) Prevalence of *Fascioloides magna* in *Galba truncatula* in the Danube backwater area east of Vienna, Austria. Wien Tierarztl Monat 98:261–267

Huňová K, Kašný M, Hampl V, Leontovyč R, Kuběna A, Mikeš L, Horák P (2012) *Radix* spp.: identification of trematode intermediate hosts in the Czech Republic. Acta Parasitol 57:273–284. doi:10.2478/s11686-012-0040-7

Kašný M, Beran L, Siegelová V, Siegel T, Leontovyč R, Beránková K, Pankrác J, Košťáková M, Horák P (2012) Geographical distribution of the giant liver fluke (*Fascioloides magna*) in the Czech Republic and potential risk of its further spread. Vet Med 57:101–109

Kennedy MJ, Acorn RC, Moraiko DT (1999) Survey of *Fascioloides magna* in farmed wapiti in Alberta. Can Vet J 40:252–254

Knapp SE, Dunkel AM, Han K, Zimmerman LA (1992) Epizootiology of fascioliasis in Montana. Vet Parasitol 42:241–246. doi:10.1016/0304-4017(92)90065-H

Krull WH (1933) A new intermediate host for *Fascioloides magna* (Bassi, 1873) Ward, 1917. Science 78:508–509. Cited in Swales WE (1935) The life cycle of *Fascioloides magna* (Bassi, 1875), the large liver fluke of ruminants, in Canada. Can J Res 12:177–215. doi:10.1139/cjr35-015

Krull WH (1934) The intermediate hosts of *Fasciola hepatica* and *Fascioloides magna* in the United States. North American Veterinary 15:13–17. Cited in Swales WE (1935) The life cycle of *Fascioloides magna* (Bassi, 1875), the large liver fluke of ruminants, in Canada. Can J Res 12:177–215. doi:10.1139/cjr35-015

Laursen JR, Stromberg BE (1993) *Fascioloides magna* intermediate snail hosts: habitat preferences and infection parameters. J Parasitol 79:302

Leontovyč R, Košťáková M, Siegelová V, Melounová K, Pankrác J, Vrbová K, Horák P, Kašný M (2014) Highland cattle and *Radix labiata*, the hosts of *Fascioloides magna*. BMC Vet Res 10:41. doi:10.1186/1746-6148-10-41

Majoros G, Sztojkov V (1994) Appearance of the large American liver fluke *Fascioloides magna* (Bassi, 1875) (Trematoda: Fasciolata) in Hungary. Parasit Hung 27:27–38

Mas-Coma S, Valero MA, Bargues MD (2009) Chapter 2. *Fasciola*, lymnaeids and human fascioliasis, with a global overview on disease transmission, epidemiology, evolutionary genetics, molecular epidemiology and control. Adv Parasitol 69:41–146. doi:10.1016/S0065-308X(09)69002-3

Novobilský A, Kašný M, Mikeš L, Kovařčík K, Koudela B (2007) Humoral immune responses during experimental infection with *Fascioloides magna* and *Fasciola hepatica* in goats. Parasitol Res 101:357–364. doi:10.1007/s00436-007-0463-5

Novobilský A, Kašný M, Pankrác J, Rondelaud D, Engström A, Höglund J (2012) *Lymnaea fuscus* (Pfeiffer, 1821) as a potential intermediate host of *Fascioloides magna* in Europe. Exp Parasitol 132:282–286. doi:10.1016/j.exppara.2012.08.005

Pfenninger M, Cordellier M, Streit B (2006) Comparing the efficacy of morphologic and DNA-based taxonomy in the freshwater gastropod genus *Radix* (Basommatophora, Pulmonata). BMC Evol Biol 6:100. doi:10.1186/1471-2148-6-100

Pybus MJ (2001) Liver flukes. In: Samuel WM, Pybus MJ, Kocan AA (eds) Parasitic diseases of wild mammals, 2nd edn. Iowa State University Press, Ames

Rajský D, Patus A, Špakulová M (1996) Rozšírenie cicavice obrovskej (*Fascioloides magna* Bassi, 1875) v jelenej chovateľskej oblasti J—I Podunajská. In: Zborník referátov a príspevkov medzinárodnej konferencie, Výskumný ústav živočíšnej výroby, Nitra, Slovakia, March 1996 (in Slovak)

Rajský D, Čorba J, Várady M, Špakulová M, Cabadaj R (2002) Control of fascioloidosis (*Fascioloides magna* Bassi, 1875) in red deer and roe deer. Helminthologia 39:67–70

Rapsch C, Dahinden T, Heinzmann D, Torgerson PR, Braun U, Deplazes P, Hurni L, Bär H, Knubben-Schweizer G (2008) An interactive map to assess the potential spread of *Lymnaea truncatula* and the free-living stages of *Fasciola hepatica*. Vet Parasitol 154:242–249. doi:10.1016/j.vetpar.2008.03.030

Rondelaud D, Novobilský A, Vignoles P, Treuil P, Koudela B, Dreyfuss G (2006) First studies on the susceptibility of *Omphiscola glabra* (Gastropoda: Lymnaeidae) from central France to *Fascioloides magna*. Parasitol Res 98:299–303. doi:10.1007/s00436-005-0067-x

Rondelaud D, Novobilský A, Höglund J, Kašný M, Pankrác J, Vignoles P, Dreyfuss G (2014) Growth rate of the intermediate snail host *Galba truncatula* influences redial development of the trematode *Fascioloides magna*. J Helminthol 88:427–433. doi:10.1017/S0022149X13000370

Sanabria R, Mouzet R, Pankrác J, Djuikwo Teukeng FF, Courtioux B, Novobilský A, Höglund J, Kašný M, Vignoles P, Dreyfuss G, Rondelaud D, Romero J (2013) *Lymnaea neotropica* and *Lymnaea viatrix*, potential intermediate hosts for *Fascioloides magna*. J Helminthol 87:494–500. doi:10.1017/S0022149X12000582

Sinitsin DF (1930) A note on the life history of the large American liver fluke, *Fasciola magna* (Bassi). Science 72:273–274. Cited in Swales WE (1935) The life cycle of *Fascioloides magna* (Bassi, 1875), the large liver fluke of ruminants, in Canada. Can J Res 12:177–215. doi:10.1139/cjr35-015

Smyth JD, Halton DW (1983) Physiology of trematodes, 2nd edn. Cambridge University Press, Cambridge

Swales WE (1935) The life cycle of *Fascioloides magna* (Bassi, 1875), the large liver fluke of ruminants, in Canada. Can J Res 12:177–215. doi:10.1139/cjr35-015

Thompson SN (1997) Physiology and biochemistry of snail-larval trematode relationship. In: Fried B, Graczyk TK (eds) Advances in trematode biology. CRC Press, Boca Raton

Vignoles P, Novobilský A, Rondelaud D, Bellet V, Treuil P, Koudela B, Dreyfuss G (2006) Cercarial production of *Fascioloides magna* in the snail *Galba truncatula* (Gastropoda: Lymnaeidae). Parasitol Res 98:462–467. doi:10.1007/s00436-005-0077-8

Vignoles P, Novobilský A, Höglund J, Kašný M, Pankrác J, Dreyfuss G, Pointier JP, Rondelaud D (2014) *Lymnaea cubensis*, an experimental intermediate host for *Fascioloides magna*. Folia Parasitol 61:185–188. doi:10.14411/fp.2014.020

Wu LY, Kingscote AA (1954) Further study on *Lymnaea stagnalis* (L.) as a snail for *Fascioloides magna* (Bassi, 1873) (Trematoda). J Parasitol 40:90–93

Chapter 5
Modern Approaches in *Fascioloides magna* Studies

Abstract The methods of cellular and molecular biology represent useful and attractive tools that have been applied in identification, taxonomy and systematics of broad spectrum of parasitic organisms over the past decades. The pilot molecular data on *Fascioloides magna* appeared in 90s of the 20th century. After more than 20 years of molecular and cellular research of *F. magna*, effective markers for accurate species identification and large-scale population studies, detailed subcellular structure of the parasite, and immunologically active molecules, were detected. This chapter is divided into four sections. First one is dealing with general structure and characterization of ribosomal genes and their utilization in molecular taxonomy and phylogeny of *F. magna*. Second part is focused on characterization and structure of mitochondrial genes and their application in studies on genetic interrelationships, biogeography, origin and transmission routes of *F. magna*. Microsatellites, biparentally inherited multilocus markers, are useful population genetics markers described in third subchapter. Data on ultrastructure, karyotype and chromosomal location of ribosomal genes of *F. magna* are presented in the last part of this chapter. In addition, we provided brief overview on current knowledge of *F. magna* isoenzyme analyses, excretory/secretory proteins, humoral immune responses during experimental infection with *F. magna* in selected final hosts, and up to date technologies of transcriptome analysis.

Keywords Giant liver fluke · Ribosomal DNA · Mitochondrial DNA · Microsatellites · Karyotype · Transcriptome · Excretory/secretory proteins · Isoenzyme analysis · Molecular taxonomy · Phylogeny

5.1 Ribosomal Genes

5.1.1 Structure and Characterization

Ribosomes are essential intracellular particles composed of proteins and RNA molecules, on which the protein synthesis is carried out (Gibbons et al. 2014). They

I. Králová-Hromadová et al., *The Giant Liver Fluke, Fascioloides magna: Past, Present and Future Research*, SpringerBriefs in Animal Sciences,
DOI 10.1007/978-3-319-29508-4_5

contain the enzymes needed to form a peptide bond between amino acids, a site for binding one mRNA molecule, and sites for bringing in and aligning the amino acids in preparation for assembly into the finished polypeptide chain (Hartl et al. 1988). Ribosomes play a key role in the process of translating the genetic information of the mRNA into protein (Prokopowich et al. 2003).

The ribosomal DNA (rDNA) sequences encoding ribosomal RNAs (rRNAs) consist of several hundred tandemly repeated copies of the transcription unit and give origin to the nucleolus (Gibbons et al. 2014). In prokaryotes, there is one to several copies of the rRNA genes. These genes may be organized in a single operon, in which they are usually separated by tRNA genes, or they may be dispersed throughout the genome (Morgan 1982). In the eukaryotic genome, ribosomal DNA represents a significant and unique type of locus (Kobayashi 2011). It contains tens to hundreds of tandemly arranged copies of the genes (mainly 30–30,000), encoding the three major rRNAs, which are an essential part of the ribosome.

Eukaryotic rDNA contains following subunits (Hillis and Dixon 1991), which are characterized in sedimentation velocity units (S, for Svedberg):

- non-transcribed spacer (NTS)
- external transcribed spacer (ETS)
- small subunit of rRNA gene (18S or SSU)
- internal transcribed spacer 1 (ITS1)
- 5.8S subunit of rRNA gene
- internal transcribed spacer 2 (ITS2)
- large subunit of rRNA gene (28S or LSU)

Ribosomal DNA represents well studied gene family; both the architecture of the rDNA and the sequence of certain domains of the genes are very highly conserved. Thus, rDNA has been an important tool for systematic studies of highly diverged taxa. On the other hand, certain regions of rDNA, which generally occur in the spacer region, are rather variable between closely related species. Therefore, they proved to be useful for species identification, as well as for phylogenetic analyses of closely related species (Collins and Paskewitz 1996).

Since different rDNA regions evolved at different rates, they can be used as genetic markers in answering questions of phylogenetic studies at many taxonomic levels (Hillis and Davis 1986; Zheng et al. 2014). The *SSU rDNA* belongs to the slowest evolving sequences found throughout living organisms; therefore it is often applied for examining ancient evolutionary events and for large phylogenetic studies. Contrary, *LSU rDNA* shows more variation in rates of evolution of its different domains which are useful for reconstructing relatively recent events. *Ribosomal spacers ITS1* and *ITS2* evolve much more rapidly. Spacer regions are characterized by high degree of variability; therefore, they can be used to infer phylogeny among closely related taxa, e.g. individuals within species (Hillis and Dixon 1991). Among molecular sequences, the ITS2 spacer belongs to one of the most frequently used markers for diagnostic and phylogenetic studies. It contains conserved secondary structure that can be used to facilitate alignments of higher taxonomic categories (from genus to order) due to its function in rRNA processing. ITS2 spacer is useful also for discrimination at the species and subspecies levels (Nei and Rooney 2005).

The reasons for the systematic versatility of rDNA include the numerous rates of evolution among different regions of rDNA (both among and within genes), the presence of many copies of most rDNA sequences per genome, and the pattern of concerted evolution that occurs among repeated copies (Hillis and Dixon 1991). In addition, the islands of highly conserved sequences within most rDNA are very useful for constructing universal primers and amplifying regions of interest by use of the PCR (Simon et al. 1990).

5.1.2 Application of Ribosomal Genes in F. magna *Studies*

The ribosomal ITS, LSU and SSU sequences were applied in molecular taxonomy of *F. magna* as species-specific markers and in phylogenetic studies (Fig. 5.1) (Bildfell et al. 2007; Králová-Hromadová et al. 2008; Lotfy et al. 2008). Table 5.1 summarises details on rDNA sequences of *F. magna* available in the GenBank.

The first reference of incomplete ITS2 sequence of *F. magna* from United States was published by Adlard et al. (1993); however, these data were not deposited in the GenBank. ITS2 region of *F. magna* was compared with respective sequence data of closely related species *Fasciola hepatica* from Australia, Hungary, Mexico and New Zealand, *Fasciola gigantica* from Indonesia and Malaysia, and *Fasciola* sp. from Japan. Intergeneric variation between *F. magna* and *F. hepatica* was 13.2 %, while 16 % variation was found between *F. magna* and *F. gigantica*. Detected variability in the nucleotide ITS2 sequence allowed molecular discrimination among species of different genera within the family Fasciolidae (Adlard et al. 1993).

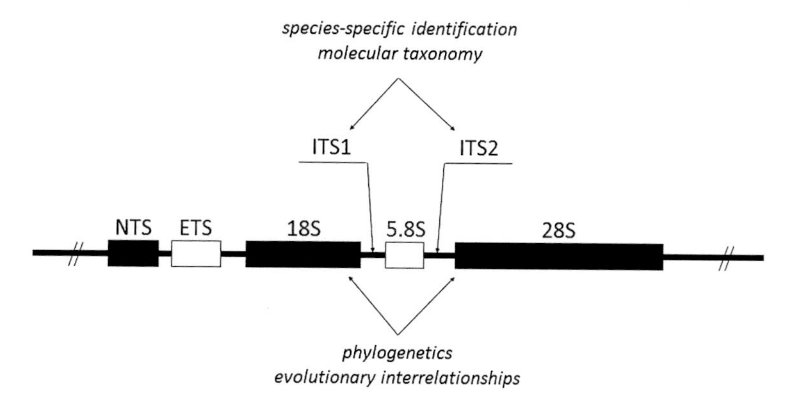

Fig. 5.1 Schematic diagram of ribosomal gene of eukaryotic DNA and application of different rDNA subunits in *F. magna* studies. Details on subunits are provided in Sect. 5.1.1

Table 5.1 GenBank submissions for ribosomal DNA sequences of *F. magna*

rDNA subunit	Accession number	Locality	References
28S	EU025872	USA/Minnesota	Lotfy et al. (2008)
18S	EF534989	Czech Republic	Králová-Hromadová et al. (2008)
18S	EF051080	USA/Oregon	Bildfell et al. (2007)
ITS1[a]	EF051080	USA/Oregon	Bildfell et al. (2007)
ITS1[a]	EF534987	Slovakia	Králová-Hromadová et al. (2008)
ITS1[a]	EF534988	Czech Republic	Králová-Hromadová et al. (2008)
ITS1[a]	EF534990	Canada/Alberta	Králová-Hromadová et al. (2008)
ITS1[a]	EF534991	USA/Oregon	Králová-Hromadová et al. (2008)
ITS1[a]	EF612475	USA/Minnesota	Lotfy et al. (2008)
ITS2	n.i.	USA	Adlard et al. (1993)
ITS2	DQ683545	Austria	Hörweg et al. (2011)
ITS2[a]	EF051080	USA/Oregon	Bildfell et al. (2007)
ITS2[a]	EF534992	Slovakia	Králová-Hromadová et al. (2008)
ITS2[a]	EF534993	Czech Republic	Králová-Hromadová et al. (2008)
ITS2[a]	EF534994	Canada/Alberta	Králová-Hromadová et al. (2008)
ITS2[a]	EF534995	USA/Oregon	Králová-Hromadová et al. (2008)
ITS2[a]	EF612487	USA/Minnesota	Lotfy et al. (2008)

[a]sequences of complete gene, *n.i.* not indicated in the respective literature

The partial sequences of the small and large rDNA subunits and complete ITS1, 5.8S, and ITS2 sequences of another US isolate (Oregon) of *F. magna* were published by Bildfell et al. (2007); these data were used for molecular identification of the parasite. Moreover, phylogenetic analysis based on SSU and ITS2 sequences revealed that *F. magna* is related to *F. hepatica* and *F. gigantica* (Bildfell et al. 2007).

Nuclear 28S rDNA, ITS1 and ITS2 spacers of *F. magna* from US (Minnesota) were obtained and analysed together with respective sequence data of six other species of fasciolids (*F. hepatica*, *F. gigantica*, *Fasciola jacksoni*, *Fasciolopsis buski*, *Protofasciola robusta* and *Parafasciolopsis fasciolaemorpha*) to study their phylogenetic interrelationships (Lotfy et al. 2008). Derived position for *F. magna* and *Fasciola* spp., an intermediate position for members of the Fasciolopsinae (*F. buski* and *P. fasciolaemorpha*), and the most basal position of *P. robusta* were revealed (Lotfy et al. 2008).

Králová-Hromadová et al. (2008) obtained almost complete 18S rDNA of *F. magna* from Czech Republic and complete ITS1 and ITS2 sequences of four isolates of *F. magna* from Slovakia, Czech Republic, Canada (Alberta) and USA (Oregon). Sequences were compared with all available and previously published sequences of different *F. hepatica* populations. The purpose of this study was to detect a level of intraspecific variation within both, *F. magna* and *F. hepatica*, as the key information for further determination of interspecific differences between both species, resulting in design of species-specific primers. While in 18S rDNA only slight interspecific sequence divergence was determined, ITS1 and ITS2 sequences provided many species-specific features for reliable discrimination of both species. Fixed interspecific genetic differences enabled to design *F. magna*-specific and *F. hepatica*-specific primers for accurate molecular identification of both species using PCR amplification or alternatively by PCR-RFLP method (Králová-Hromadová et al. 2008). Later on, the study was extended in newly designed ITS2 species-specific primers for two other veterinary important gastrointestinal trematodes of ruminants, *Paramphistomum cervi* and *Dicrocoelium dendriticum* (Bazsalovicsová et al. 2010). All four species-specific primers were applied in the genotypization of morphologically hardly distinguishable eggs of respective flukes, what is important for in vivo coprological diagnostics (Oberhauserová et al. 2010).

Fascioloides magna-specific ITS2 primers, designed by Králová-Hromadová et al. (2008), were used for molecular identification of novel findings of giant liver fluke in several countries. Genotypization of young or undifferentiated rediae in intermediate snail hosts from Danube backwater in Austria confirmed presence of *F. magna* (Hörweg et al. 2011). In the Lower Silesian Wilderness (southwestern Poland), *F. magna*-specific ITS2 primers were used for determination of trematode eggs found in faecal samples collected from red deer (Pyziel et al. 2014). The study corroborated presence of *F. magna* in southwestern Poland after more than 60 years. In another Polish locality, Podkarpackie Province in southeastern Poland, adult flukes were found after necropsy of fallow deer (Karamon et al. 2015). Their genotypization with *F. magna*-specific ITS2 primers validated presence of giant liver fluke (Karamon et al. 2015).

5.2 Mitochondrial Genes

5.2.1 Structure and Characterization

Mitochondrial markers have proved to be useful for inferring patterns of population genetic structure in many parasitic organisms (Nadler et al. 1995). Mitochondria, as semi-autonomous organelles with own mitochondrial genome, can replicate independently of the nuclear genome; this process is called autoreplication (Crimi and Rigolio 2008). The mitochondrial genome is formed by mitochondrial DNA (mtDNA) localized in mitochondrial matrix and is responsible for extranuclear inheritance.

Mitochondrial DNA is formed by circular double-stranded molecule, which is thought to be strictly maternally inherited (Birky 2001). Multiple copies of mtDNA facilitate its isolation and amplification. The genetic function of mtDNA is well-conserved and involves five mitochondrial processes: respiration and/or oxidative phosphorylation, transcription, translation, RNA maturation and protein import (Burger et al. 2003). Through a series of enzymatic processes of oxidative phosphorylation mitochondria supply cells with the energy-rich ATP molecules (Crimi and Rigolio 2008).

Mitochondrial DNA is a relatively small, abundant and easy to isolate DNA molecule, which consists of the following genes:

- 2 ribosomal RNA genes (*rrn*S, *rrn*L), coding components of mitochondrial ribosomes
- 22 transfer RNA genes (*trn*), required in the translation process of mitochondrial proteins
- 12–13 protein coding genes, coding following enzymes of oxidative phosphorylation:
  - cytochrome *c* oxidase complex (*cox*1–*cox*3; 3 subunits)
  - cytochrome *b* (*cob*)
  - adenosine triphosphatase complex (*atp*6, alternatively *atp*8)
  - nicotinamide dehydrogenase complex (*nad*1–*nad*6 and *nad*4L; 7 subunits)

Gene products of small (*rrn*S) and large (*rrn*L) subunits of *rrn* mitochondrial gene form the mitochondrial ribosomes (Noller 1991). Twenty-two *trn* genes are scattered throughout the mitochondrial genome. The arrangement of protein-coding genes, ribosomal RNA genes and transfer RNA genes are conserved (Le et al. 2000). The structure and gene arrangement of mitochondrial DNA of *F. hepatica* is illustrated in Fig. 5.2.

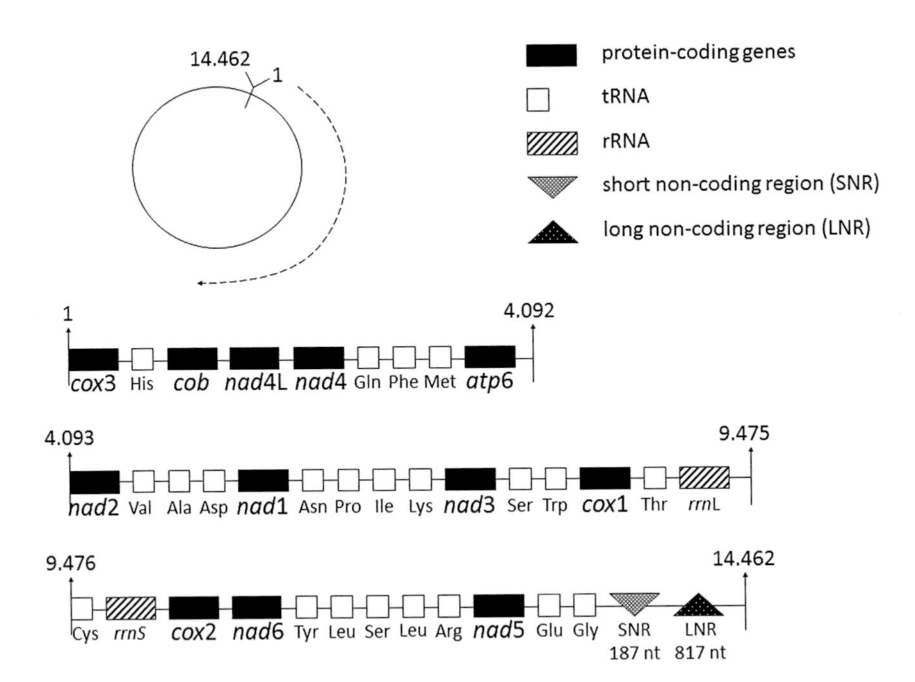

Fig. 5.2 Schematic diagram of complete mitochondrial genome of *F. hepatica* (complete genome published by Le et al. 2000)

The mtDNA sequences of animals evolve faster than nuclear genes; therefore they are suitable for detection of interrelationships among closely related organisms (Wolstenholme 1992; Avise 1994; Boore 1999). The high frequency of mutations, absence of recombination and hybridisation, approximately constant content of genes and variation in gene codes are unique characteristics of mtDNA (Hu et al. 2004), which predestinates it to be suitable molecular marker for population genetics, phylogeny, and studies of biogeography/phylogeography (Avise and Walker 1999).

5.2.2 *Application of Mitochondrial Genes in* F. magna *Studies*

Mitochondrial genome has been sequenced for a broad spectrum of organisms, including an increasing number of parasites (Feagin 2000). The first complete mitochondrial genome of parasitic worm was that of *Ascaris suum* (Nematoda) (Okimoto et al. 1992); out of Platyhelminthes, the first complete mtDNA was acquired for medically important tapeworm *Echinococcus multilocularis* (Le et al. 2000). Currently, over 70 submissions of complete mitochondrial genomes of flukes (Trematoda) are available in the GenBank (http://www.ncbi.nlm.nih.gov/nuccore; 2015 December 15th). Out of them, two submissions belong to closely related species *Fasciola hepatica* (AF216697, NC002546, Le et al. 2000, 2001). The complete mitochondrial genome for *F. magna* is not available yet. Complete or partial sequences of most frequently studied mitochondrial genes of *F. magna*, in particular cytochrome *c* oxidase subunit I (*cox*1) and nicotinamide dehydrogenase subunit I (*nad*1) are summarized in Tables 5.2 and 5.3.

Mitochondrial genes as tools for assessment of genetic interrelationships Apart from other characteristics, such as veterinary importance and broad host spectrum, *F. magna* represents a very remarkable species also due to its large spatial distribution, invasive character, and potential to colonize new territories. These features predestined *F. magna* to be a good model for elucidation of genetic interrelationships between and within North American and European populations and assessment of an origin of European populations. For that purpose, suitable molecular, in particular mitochondrial markers, had to be designed.

As a starting point, complete sequences of *cox*1 and *nad*1 for representatives of three allopatric populations of *F. magna* coming from Slovakia, Czech Republic and United States (Oregon) were obtained by Králová-Hromadová et al. (2008) (Fig. 5.3). Their comparison revealed 28 and 13 point mutations, respectively. This allowed the selection of shorter variable regions (*cox*1, 384 bp; *nad*1, 405 bp) which displayed high level of sequence divergence and were proposed to be applied as effective markers for further population studies on *F. magna* (Králová-Hromadová et al. 2008).

Table 5.2 GenBank submissions for mitochondrial *cox*1 sequences of *F. magna*

Accession no.	Locality	R	Accession no.	Locality	R
EF534996[a]	Slovakia	1	KP635013	Canada/B. Columbia	3
EF534997[a]	Czech Republic	1	KP635014	Canada/Labrador	3
EF534998[a]	USA/Oregon	1	KP635015	Canada/Labrador	3
GU599861	Canada/Alberta	2	KP635016	Canada/Labrador	3
GU599871	Canada/Alberta	2	KP635017	Canada/Quebec	3
GU599862	USA/Oregon	2	KP635018	Canada/Quebec	3
GU599873	USA/Minnesota	2	KP635019	Canada/Quebec	3
GU599874	USA/Minnesota	2	KP635020	Canada/Quebec	3
GU599875	USA/Minnesota	2	KP635021	USA/Minnesota	3
GU599876	USA/Minnesota	2	KP635022	USA/Mississippi	3
GU599877	USA/Mississippi	2	KP635023	USA/Georgia	3
GU599878	USA/Mississippi	2	KP635024	USA/Louisiana	3
GU599879	USA/Mississippi	2	KP635025	USA/Louisiana	3
GU599880	USA/Mississippi	2	KP635026	USA/Louisiana	3
GU599882	USA/Florida	2	KP635027	USA/Louisiana	3
GU599872	USA/Georgia	2	KP635028	USA/Louisiana	3
GU599881	USA/Louisiana	2	KP635029	USA/Louisiana	3
GU599860	Italy	2	KP635030	USA/South Carolina	3
GU599863	Italy	2	KP635031	USA/South Carolina	3
GU599870	Czech Republic	2	KP635032	USA/South Carolina	3
GU599868	Czech Republic	2	KP635033	USA/South Carolina	3
GU599864	Czech Republic	2	KP635034	USA/South Carolina	3
GU599869	Slovakia	2	KP635035	USA/South Carolina	3
GU599865	Slovakia	2	KP635036	USA/South Carolina	3
GU599866	Hungary	2	KP635008	Poland	4
GU599867	Croatia	2	KF784787	Austria	5
KP635011	Canada/B. Columbia	3	KF784788	Austria	5
KP635012	Canada/B. Columbia	3			

[a]sequences of complete gene, *B. Columbia* British Columbia, *R* reference, *1* Králová-Hromadová et al. (2008), *2* Králová-Hromadová et al. (2011), *3* Bazsalovicsová et al. (2015), *4* Králová-Hromadová et al. (2015), *5* Sattmann et al. (2014)

The selected variable *cox*1 and *nad*1 fragments were applied in order to reveal an origin of European *F. magna* populations, the course of colonisation and migratory routes of this alien parasite in Europe (Králová-Hromadová et al. 2011). An extensive material of *F. magna* populations coming from all European natural foci, Italy (IT), Czech Republic (CZ) and Danube floodplain forests (DFF), and comparative samples from North America were studied.

The principal outcome of the study was determination of two independent phylogenetic lineages of *F. magna* from Europe. The Italian population represented one distinct phylogenetic clade, while the second clade included populations from

Table 5.3 GenBank submissions for mitochondrial *nad*1 sequences of *F. magna*

Accession no.	Locality	R	Accession no.	Locality	R
EF534999[a]	Slovakia	1	KP635038	Canada/B. Columbia	3
EF535000[a]	Czech Republic	1	KP635039	Canada/B. Columbia	3
EF535001[a]	USA/Oregon	1	KP635040	Canada/B. Columbia	3
GU599845	Canada/Alberta	2	KP635041	Canada/Quebec	3
GU599846	Canada/Alberta	2	KP635042	Canada/Quebec	3
GU599848	USA/Oregon	2	KP635043	Canada/Quebec	3
GU599849	USA/Minnesota	2	KP635044	Canada/Quebec	3
GU599850	USA/Minnesota	2	KP635045	Canada/Quebec	3
GU599851	USA/Minnesota	2	KP635046	Canada/Quebec	3
GU599852	USA/Minnesota	2	KP635047	Canada/Labrador	3
GU599855	USA/Mississippi	2	KP635048	Canada/Labrador	3
GU599856	USA/Mississippi	2	KP635049	Canada/Labrador	3
GU599857	USA/Mississippi	2	KP635050	USA/Minnesota	3
GU599853	USA/Florida	2	KP635051	USA/Mississippi	3
GU599854	USA/Florida	2	KP635052	USA/Florida	3
GU599847	USA/Georgia	2	KP635053	USA/Florida	3
GU599858	USA/Louisiana	2	KP635054	USA/Georgia	3
GU599859	USA/Louisiana	2	KP635055	USA/Louisiana	3
GU599834	Italy	2	KP635056	USA/Louisiana	3
GU599835	Italy	2	KP635057	USA/Louisiana	3
GU599836	Italy	2	KP635058	USA/South Carolina	3
GU599837	Czech Republic	2	KP635059	USA/South Carolina	3
GU599838	Czech Republic	2	KP635060	USA/South Carolina	3
GU599839	Czech Republic	2	KP635061	USA/South Carolina	3
GU599840	Czech Republic	2	KP635062	USA/South Carolina	3
GU599841	Slovakia	2	KP635063	USA/South Carolina	3
GU599842	Slovakia	2	EF612499	USA/Minnesota	4
GU599843	Hungary	2	KP635009	Poland	5
GU599844	Croatia	2	KF784789	Austria	6
KP635037	Canada/B. Columbia	3	KF784790	Austria	6

[a]sequences of complete gene, *B. Columbia* British Columbia, *R* reference, *1* Králová-Hromadová et al. (2008), *2* Králová-Hromadová et al. (2011), *3* Bazsalovicsová et al. (2015), *4* Lotfy et al. (2008), *5* Králová-Hromadová et al. (2015), *6* Sattmann et al. (2014)

CZ and DFF, what clearly indicated that *F. magna* was introduced to Europe at least twice (Králová-Hromadová et al. 2011). Molecular data did not show any genetic relationships between flukes from Italy and other European foci; it was apparent that *F. magna* in Italian La Mandria did not spread further to Europe and remained isolated (Králová-Hromadová et al. 2011).

Since Italian population of *F. magna* clustered with specimens from Alberta (Canada) and Oregon (USA), a western North American origin of Italian focus was

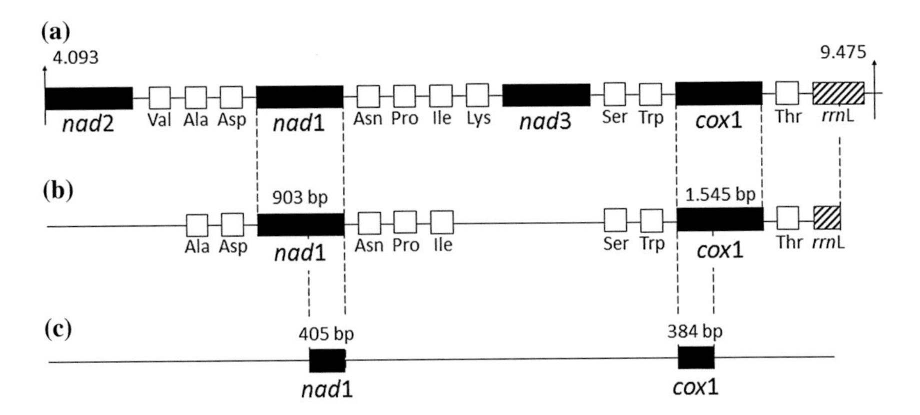

Fig. 5.3 **a**—selected part of complete mitochondrial genome of *F. hepatica* (Le et al. 2000), **b**—sequences of complete *nad*1 and *cox*1 genes and adjacent regions of *F. magna* acquired by Králová-Hromadová et al. (2008), **c**—short variable regions of selected mitochondrial genes of *F. magna* designed by Králová-Hromadová et al. (2008) and applied in population studies of European (Králová-Hromadová et al. 2011) and North American (Bazsalovicsová et al. 2015) populations

confirmed. On the other hand, flukes from CZ and DFF displayed close genetic relationships with *F. magna* from southeastern USA (Fig. 5.4) (Králová-Hromadová et al. 2011).

Concerning the third and the most recent European focus, Danube floodplain forests, Slovak, Hungarian and Croatian *F. magna* samples represented the same genetic pool. It was evident that similar ecological conditions in the floodplain forests down the Danube River provide an excellent natural environment for intermediate and final hosts of *F. magna* and facilitate its further spread down the river. Identical haplotypes detected for parasites from CZ and DFF implied that *F. magna* was introduced to the Danube region from already established Czech focus, although the means of transfer remained unresolved (Králová-Hromadová et al. 2011).

In North America, population and genetic structure of giant liver fluke was studied by Bazsalovicsová et al. (2015) using the same *cox*1 and *nad*1 fragments as designed by Králová-Hromadová et al. (2008). The principal outcome was detection of two separate lineages of *F. magna* in North American continent. The western lineage was formed by individuals from Rocky Mountain trench (RMT) (Alberta, Canada) and northern Pacific coast (NPC) (British Columbia, Canada and Oregon, USA) whereas the eastern lineage was created by samples from the Great Lakes region (GLR) (Minnesota, USA), Gulf coast, lower Mississippi, and southern Atlantic seaboard region (SAS) (Mississippi, Louisiana, South Carolina, Georgia, Florida) and northern Quebec and Labrador (NQL, Canada) (Fig. 5.4). Results of molecular analyses were discussed with historical and current distribution of final cervid hosts of *F. magna*. It was assumed that genetic separation of two mitochondrial lineages very probably correlates with data on

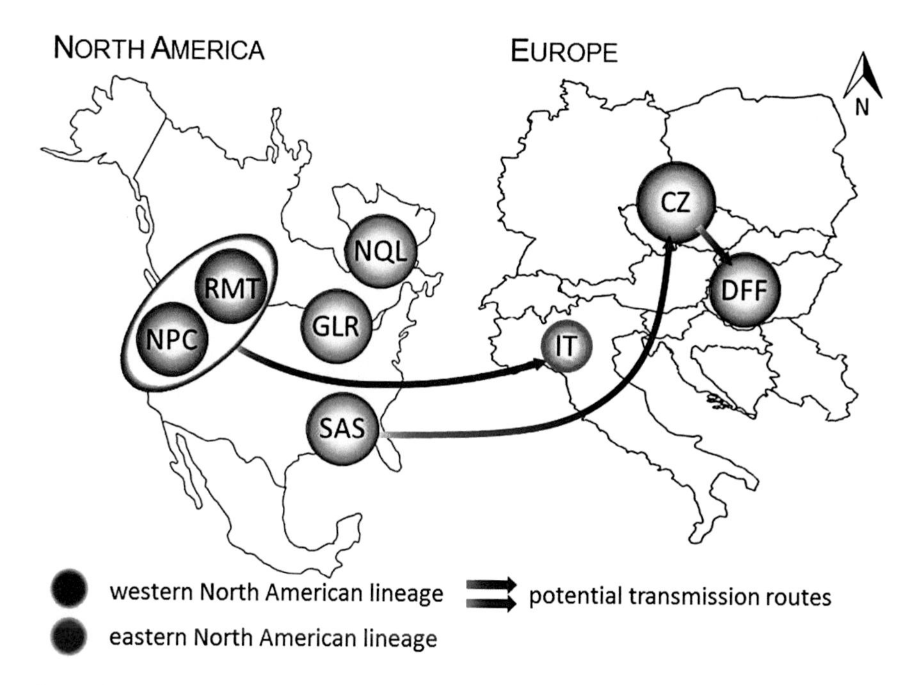

Fig. 5.4 Interrelationships of *F. magna* from North America and Europe revealed by mitochondrial data (Králová-Hromadová et al. 2011; Bazsalovicsová et al. 2015)

historical distribution of white-tailed deer in eastern and wapiti in western part of North America. Since haplotypes determined in *F. magna* representatives were shared amongst several adjacent populations, no signs of host specificity of the parasite towards any cervid species was noticeable (Bazsalovicsová et al. 2015).

High Resolution Melting, an effective screening method The short variable regions within *cox*1 and *nad*1 genes, as designed by Králová-Hromadová et al. (2008), were proved to be suitable for large-scale populations studies and assessment of genetic interrelationships of geographically distinct populations of *F. magna* (Králová-Hromadová et al. 2011). The future application of these markers in extensive data-sets would require financially demanding and time consuming experiments. Therefore, an implementation of fast and effective screening method has arisen as a necessary need.

The High Resolution Melting method (HRM) is a screening tool, which was established as a single tube method for single nucleotide polymorphism (SNP) genotyping and detection of unknown sequence variants (Gundry et al. 2003; Wittwer et al. 2003). The principle of HRM method is denaturation of double stranded (ds) DNA template by heat (up to 95 °C) what results in single strand (ss) DNA. Fluorescent dyes applied during HRM analysis have higher level of fluorescence when bound to dsDNA than in ssDNA bound or in unbound state (Ririe et al. 1997; Wittwer et al. 2003). The result of HRM is a melting curve that can be generated if the fluorescence is continuously monitored during the heating of a sample through its dissociation temperature (melting temperature, *T*m). Each DNA template has specific *T*m depending on the nucleotide structure and is characterized by the specific melting curve. Detection and differentiation of different sequence variants rely on change in the *T*m and shape of the melting curve. Samples containing the same sequence variants are identified as groups that exhibit similar melting profiles (Bennett et al. 2003).

Previously selected variable region of mitochondrial *cox*1 was applied in development and validation of HRM method for studies of European *F. magna* populations (Radvánský et al. 2011). Three internal nucleotide probes were originally designed and used for optimization and reliable discrimination of haplotype-specific melting curves for unique haplogroups, assorted previously according to the sequence analysis. The successfully optimized HRM method provided the opportunity for rapid and effective "close tube" analysis of novel haplotypes in future studies of *F. magna* from Europe (Radvánský et al. 2011).

The HRM method was consequently applied to determine an origin of *F. magna* from Croatia (Bazsalovicsová et al. 2013). Two structurally different *cox*1 haplotypes were detected for 200 *F. magna* individuals from the Baranja region (northeastern Croatia) employing three internal oligonucleotide probes (Radvánský et al. 2011). Both haplotypes were easily distinguishable by specific melting peaks identical to the reference haplotypes (Ha3 and Ha4; Králová-Hromadová et al. 2011) which revealed their Danube origin. The results confirmed that Danube floodplain forests represents expanding natural focus and a spread of fascioloidosis further down the Danube River is very probable (Bazsalovicsová et al. 2013).

Mitochondrial genes as tools for molecular taxonomy and phylogeny Although characteristics of mitochondrial genes predestine them to be applied mainly in population genetic studies, they can serve also as markers utilized in molecular taxonomy and phylogeny. For elucidation of poorly known patterns of diversification, origin and biogeography of fasciolids, partial *nad*1 gene (488 bp) of mtDNA was applied (Lotfy et al. 2008). Evolutionary relationships among fasciolids revealed that *F. magna* is monotypic and grouped closely with *Fasciola jacksoni*, while three other *Fasciola* species were detected to be paraphyletic (Lotfy et al. 2008).

The partial *cox*1 and *nad*1 sequences were also obtained for two *F. magna* isolates (adult fluke from red deer and rediae from *G. truncatula*) from Austria (Sattmann et al. 2014). Sequences displayed 100 % identity to each other, and they were compared with respective *F. magna*-specific data available in the GenBank as determined by Králová-Hromadová et al. (2011). Partial *cox*1 gene was identical to *F. magna* from Czech Republic, Slovakia, Hungary and Croatia; *nad*1 showed 100 % identity to isolates from Czech Republic and Slovakia. The results indicated very close relation of *F. magna* from Austria to populations of the parasite from neighbouring countries (Sattmann et al. 2014).

The genetic structure of *F. magna* individuals recently found in Poland (Lower Silesian Wilderness) (Pyziel et al. 2014) was determined using mitochondrial *cox*1 and *nad*1 molecular markers (Králová-Hromadová et al. 2015). Sequence data were compared with the respective molecular data available for all European populations of *F. magna* (Králová-Hromadová et al. 2011). The study revealed the genetic uniformity of *F. magna* specimens from Poland with individuals from the Czech focus. Based on these findings, it was evident that Czech focus was enlarged and the description of the second European focus of *F. magna* was suggested to be modified as "the Czech Republic and southwestern Poland" (Králová-Hromadová et al. 2015).

5.3 Microsatellites

5.3.1 Structure and Characterization

In last decade, microsatellites (*syn.* STR, Short Tandem Repeats; or SSR, Simple Sequence Repeats) have been developed as one of the most popular classes of genetic markers owing to their high reproducibility, multi-allelic nature, codominant mode of inheritance, abundance and wide genome coverage (Schlötterer 2004). Genetic structuring of natural populations is routinely investigated using microsatellite markers (Balloux and Lugon-Moulin 2002), but they proved to be also suitable molecular markers for population genetic studies, diagnostics, species identification, and phylogeographical analysis of higher taxonomical units. Microsatellites are ideal markers for studies of biodiversity, paternity testing, determination of interrelationships between closely related species, and migrations of populations. Considerable polymorphism in STR may be characterized due to variation in the number of repeat units. However, microsatellite polymorphism is sufficiently stable to use in genetic analyses (Hearne et al. 1992).

Microsatellites are scattered throughout a broad spectrum of prokaryotic and eukaryotic genomes in multiple copies, both in protein-coding and non-coding regions (Zane et al. 2002). Despite their ubiquitous occurrence, microsatellite density and distribution vary markedly across genomes (Dieringer and Schlötterer 2003). They consist of tandemly repeated sequences of very short nucleotide motifs (1–6 bp) with possible different functional roles (Buschiazzo and Gemmell 2006). Microsatellite loci with ten or more repeats are particularly likely to have many alleles (Queller et al. 1993).

The genomic distribution of microsatellites is determined by the repeat type (mono-, di-, tri-, tetranucleotide etc.) and the sequence of the repeat. The number of repeats of nucleotide motifs may differ in each locus; there is a high polymorphism in number of repeats per locus in natural populations connected with high level of variability. The mutuality in case of STR is around 0.1 % mutations per generation (Flegr 2009). There are three types of STR (Fig. 5.5); any kind of its combination is possible.

The presence of high number of STR loci in genome may be potential important tool for analyses in almost any problem requiring Mendelian markers (Queller et al. 1993). Since they are usually less than 100–150 bp long and are localized in DNA

Types of microsatellite repeats

...CACACACACACACACACACACACA...	perfect/pure $(CA)_{12}$
...CACACACACATGTGTGCACACACACACACACACA...	interrupted/imperfect $(CA)_5(TG)_3(CA)_9$
...CACACACATGCTGCTGCCACACACATGCTGCTGC...	compound $(CA)_4(TGC)_3(CA)_4(TGC)_3$

Fig. 5.5 Schematic diagram of structure and arrangement of microsatellite motifs

region with unique sequences, they can be amplified for identification by PCR (Hearne et al. 1992; Flegr 2009).

In newly analysed taxa, microsatellites need to be de novo isolated (Zane et al. 2002). The procedures applied in the STR design have developed according to the technical and methodological possibilities. The "traditional" cloning method (cloning of genomic DNA fragments enriched for STRs), used for isolation of microsatellite clones, was quite time consuming and involved several steps. It was necessary to create a small insert, to perform partial genomic library in a plasmid or phage vector, and then screen clones by repeated rounds of filter hybridization using an oligonucleotide repeat probe. In order to increase the proportion of clones in a given library containing the STR motif of interest, microsatellite enrichment has been developed (Nunome et al. 2006; Andrés and Bogdanowicz 2011). However, traditional strategies approved to be less useful and inefficient for species with low microsatellite frequencies (Zane et al. 2002 and references therein).

High-throughput next-generation sequencing (NGS) technologies are revolutionizing the field of evolutionary biology; they obviously supplant traditional cloning methods for the discovery of microsatellite loci and opened the new possibilities for genetic analysis at different levels (Andrés and Bogdanowicz 2011). Due to decreasing cost of NGS technologies, such as pyrosequencing (e.g. 454 sequencing) and sequencing by synthesis (e.g. Illumina), genome-wide evolutionary questions using hundreds of individuals for many organisms is possible at fraction of the cost and effort of traditional approaches (Etter et al. 2011; Zalapa et al. 2012). The main benefit of these technologies is avoiding the need for cloning of DNA fragments, since sequence data are determined from amplified single DNA fragments (Ansorge 2009).

The NGS approaches generate large amount of the sequencing data from microsatellite enriched libraries or genomic DNA which are then mined for microsatellite loci (typically thousands). Primers are designed for the regions flanking the microsatellite repeat and then are tested to identify markers with consistent PCR amplification of unique polymorphic loci. Co-amplification of multiple microsatellites in a single cocktail (multiplexing) makes a procedure much easier, faster and cheaper (Guichoux et al. 2011).

5.3.2 *Design and Future Application of Microsatellites in* F. magna *Studies*

Recently, next-generation sequencing approach was used to develop multiplex panels and performed de novo design of primers for microsatellite loci in *F. magna* (Minárik et al. 2014). Since the data on STR markers of giant liver fluke were neither available in the literature, nor in the GenBank databases, microsatellites were originally designed using following methodological steps:

- microsatellite mining using selective library enrichment and next-generation sequencing for identification of specific repetitive motifs
- selection of suitable STR candidates designed after NGS analysis by several tests:
  - detection of amplification effectiveness of designed PCR primers
  - confirmation of the presence of declared repetitive motif by Sanger sequencing
  - determination of STR allele polymorphism of tested loci by fragment analysis with fluorescently labelled primers by capillary electrophoresis
- development of multiplex panels
- statistical analyses of selected loci for observed (Ho) and expected (He) heterozygosity, deviations from Hardy-Weinberg equilibrium (HWE) and the presence of null alleles (Fig. 5.6).

Out of 667 amplicon candidates generated after NGS, 118 provided the best resolution and were tested with PCR analyses. In total, 56 yielded PCR products of expected size, and in 36 of them the declared repetitive motif was identified by Sanger sequencing. Finally, 11 microsatellite loci were recommended for future use

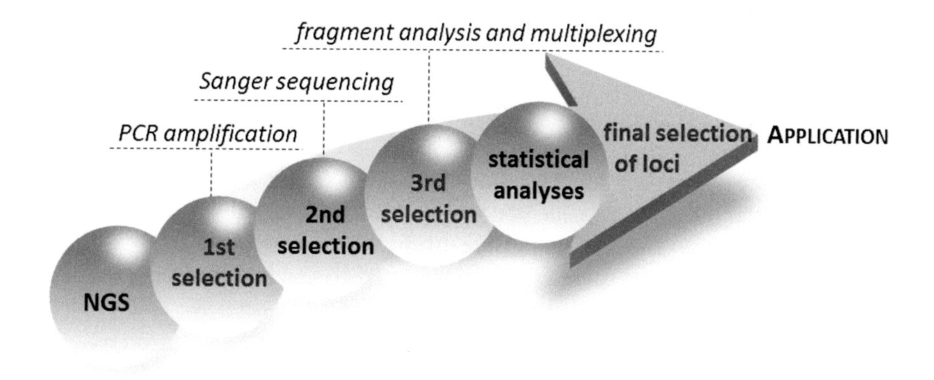

Fig. 5.6 Methodological steps of design of microsatellite markers for *F. magna* (Minárik et al. 2014)

in studies of genetic interrelationships of North American and European populations of *F. magna* (Minárik et al. 2014).

The multilocus approach based on the analyses of 11 microsatellite loci has been currently applying in order to resolve the patterns of population genetic structure and admixture of *F. magna* from Europe and North America. Further aims of the study are to determine the exact origin of European populations of the parasite, and to reveal its potential transmission and migratory routes in North America and Europe (Juhásová et al. unpublished).

5.4 Other Molecular, Cellular and Immunological Studies of *F. magna*

5.4.1 Ultrastructure

Microscopy has dramatically expanded our horizonts in the field of helminthology by providing fundamentally important information on the structure and functional correlates of a number of key organ systems of parasitic organisms. Many important discoveries have been accrued by way of transmission and scanning electron microscopy, as well as from use of the confocal scanning laser microscope. The techniques provide valuable investigative tools not only in taxonomic and structural studies by revealing inter- and intraspecific variations and regional differences in surface topography, but also in understanding host-parasite relationships and evaluation of drug effects (e.g. Halton 2004).

Using the scanning electron microscopy, the surface features of adult *F. magna* from USA were described (Naem et al. 2012). The dorsoventrally flattened body was covered by spiny tegument, which was slightly different in ventral and dorsal side. At the ventral side, well developed spines and small ciliated papillae (anterior end), spines with two or more points and dome-shaped papillae (mid-region), and small spines with blunted edges and few papillae (posterior end), were observed. At the anterior end of the dorsal side, small dome-shaped and ciliated papillae were present; more prominent spines than the papillae were typical for mid-region of the dorsal side. At the posterior end of the dorsal side, spines became progressively fewer, smaller and shorter. Patterned tegument characterized by dome-shaped and ciliated papillae was typical for the surface of the oral sucker, while ventral sucker possessed a smooth surface with two spine-like structures. Both suckers were surrounded by spines; some of them were small with a sharp point and some had serrated edges. Contrary to spine free-areas observed around the excretory pore, the surface of the cirrus organ showed small groups of tiny spines in some areas (Naem et al. 2012).

5.4.2 Karyotype

Chromosomes are hereditary elements of the whole nuclear genome. Karyological features may indicate the evolutionary distance between animals of different taxonomic categories that may not be obvious at morphological level (Dobigny et al. 2004). Cytogenetic studies can provide basic data such as chromosome number and classical karyotype features (banding patterns, karyotype asymmetry, secondary constrictions), as well as detailed information on location of genes on chromosomes by fluorescence in situ hybridization (FISH) (e.g. Bombarová et al. 2009, 2014).

An important progress in the research of karyotypes was made by an improvement of banding methods, which offered much more detailed and precise analysis of inner structure of chromosomes. Location of a secondary constriction (i.e. nucleolar organizer region, NOR) is frequently used as a marker for cytotaxonomy and phylogenetic comparisons in invertebrates (e.g. Bombarová et al. 2007; Nguyen et al. 2010). Comparative karyotype analyses and chromosome data mapping can become a powerful tool for evolutionary and taxonomic studies.

The original description of chromosome set of *F. magna* was based on mitotic divisions of spermatogonial cells isolated from fluke testes (Reblánová et al. 2010). The cytogenetic study revealed that the karyotype of giant liver fluke comprises 11 pairs of medium-sized chromosomes, which were relatively similar in their length and morphology. The karyotype formula was ascertained as 2n = 22, n = 9 st + 1sm-m + 1 sm. The total length of the complement (TCL) reached 35.17 μm; the first longest pair measured 4.65 μm (13.25 % of TCL) and the absolute length decreased to the 1.92 μm (5.43 % of TCL) of the last chromosome pair. Rather small amount of heterochromatin distributed predominantly near the centromeric region of all 11 chromosome pairs was observed after fluorescent DAPI-staining. Moreover, DAPI-positive bands were detected at chromosomes No. 5 (at the end of the long arms) and No. 6 (interstitially on the long arms) (Reblánová et al. 2010).

In order to visualize localization of ribosomal genes on the chromosomes of *F. magna*, the FISH technique with 18S rDNA probe was performed (Reblánová et al. 2011). The study revealed a single cluster of ribosomal genes in mitotically dividing spermatogonia. Chromosomal location of ribosomal loci was found to be situated interstitially, in pericentromeric regions of the long arms of the submetacentric pair No. 10. Out of all hermaphroditic trematodes, *F. magna* together with closely related fluke *F. hepatica* were the first model species studied by FISH technique (Reblánová et al. 2011).

5.4.3 Isoenzymes

Isoenzyme (or isozyme) is the term applied to electrophoretically distinct enzymatic proteins, separated primarily on the bases of differences in net charge. Over the past 40 years, isoenzyme approaches have been widely used in studies of parasite systematics in order to distinguish between morphologically similar species. Isoenzyme analyses have provided numerous genetic markers relating to population structure and gene flow among populations (e.g. Nadler and De Leon 2011). Although the technique has been commonly applied to provide answers for taxonomic studies it remains underutilized, perhaps because of recent advances in modern DNA-based molecular technologies (Andrews and Chilton 1999).

Using isoenzyme analysis, Lydeard et al. (1989) examined spatial genetic differentiation in *F. magna* populations collected from white-tailed deer from four sampling sites in the southeastern United States (Tennessee, Kentucky, South Carolina). Nine of 14 resolved electrophoretic loci (64 %) were detected as polymorphic. Significant allele frequency differences among samples were determined for seven out of nine loci. Genetic distance values increased with geographic distance among samples. The low degree of genetic divergence between localities within a state suggested that some gene flow may occur among populations.

In a similarly designed population study, genetic structuring in *F. magna* and its definitive hosts (white-tailed deer) was studied by Mulvey et al. (1991) in hunting areas in South Carolina. Five polymorphic loci were used to estimate genetic variation, and four of them had significant heterozygote deficiencies over the entire examined territory. The study revealed significant spatial genetic differentiation for flukes and deer hosts that may be partially due to the complex life cycle of the flukes. Patterns of genetic distance in *F. magna* individuals were not congruent with those of cervid hosts nor were they correlated with geographic distance between locations. Spatial genetic differentiation among flukes was influenced by the aggregated distribution of flukes in hosts and their asexual reproduction in the snail that leads to the release of multiple individuals from a limited number of clones.

5.4.4 *Immunology*

Immunoassays have indispensable role in the highly sensitive qualitative and quantitative detection within heterogeneous samples for over 50 years. In research (including parasitology), industry and medical practice, the most useful and sensitive methods are immunoblot (Western blot) and enzyme-linked immunosorbent assay (ELISA), which are typically used for the detection of specific proteins (Tighe et al. 2015). Western blotting analysis can detect the presence of a specific protein in a solution that contains a number of proteins, according to their molecular weight (Yang and Ma 2009). In contrast, ELISA method can accurately quantitate intracellular or extracellular proteins using an enzyme to detect the binding of antigen and/or antibody (Tighe et al. 2015). Using immunoassays, parasites can be identified cost effectively and in a timely manner (Josko 2012).

The above techniques were also used to evaluate humoral immune response and dynamics of antibodies in goats experimentally infected with *F. magna* (Novobilský et al. 2007). The ELISA test was used for determination of serum antibody responses after *F. magna* infection. The significant increases of specific antibodies against *F. magna* excretory/secretory products (FmESP) were observed in all infected goats since two weeks post infection. Besides, the cross-reaction of antibodies against *F. magna* and closely related *F. hepatica* with ESP proteins was recorded, what resulted in conclusion that the method is highly sensitive but not specific. The antigenic properties of FmESP and FhESP were in parallel characterized by immunoblot analysis. Based on the results, 40 and 120 kDa species-specific proteins of *F. magna*, and 80 and 160 kDa proteins of *F. hepatica* were suggested as the most suitable candidates, since no cross reaction was observed (Novobilský et al. 2007).

5.4.5 Transcriptome and Excretory/Secretory Proteome

Over the past decade, the next-generation sequencing (NGS) technologies have greatly accelerated our understanding of the complexity of gene structure, expression, and regulation (He et al. 2012). Novel DNA sequencing systems (e.g. Roche's 454 Genome sequencer, Illumina's Solexa IG sequencer, Applied Biosystem's SOLiD system, etc.) provide new opportunities of rapid and reliable ways for: (i) searching of genetic variants by sequencing the whole genome or targeted genome regions; (ii) profiling of mRNAs, small RNAs; (iii) characterization of transcriptomes and proteomes, as well as precise analysis of RNA transcripts for gene expression; (iv) identification of DNA regions that interact with regulatory proteins in functional regulation of gene expression (e.g. Ansorge 2009).

In *F. magna* individuals, the transcriptome and secreted proteome were originally characterized using the Illumina sequencing technology, one-dimensional SDS-PAGE and OFFGEL protein electrophoresis (Cantacessi et al. 2012). In total, 20,140 peptides were inferred from the transcriptome of *F. magna* which were classified based on homology searches, protein motifs, gene ontology and biological pathway motifs and consequently assigned to biological process, cellular component and molecular function. Molecules containing a predicted signal peptide (835) were most abundant in the transcriptome; out of them 80 were identified in the excretory/secretory products. Antioxidant proteins, followed by peptidases and proteins involved in carbohydrate metabolism, were also highly represented.

In addition, analysis of transcriptome revealed 42 transcripts encoding cathepsins (especially cathepsin B and L) of which eight were identified in the excretory/secretory products and seven were predicted to contain a signal peptide indicative of secretion (Cantacessi et al. 2012). From the proteome predicted for the giant liver fluke, 48.1 % (9,690) proteins were assigned to 384 biological pathway terms such as spliceosome, RNA transport and endocytosis. In the excretory/secretory products, cystatin (cysteine protease inhibitor) was found to be the most abundant, followed by two cathepsin L1 proteases, cathepsin B, calpain, ferritin and 12 lysosomal proteins. Cathepsin L with four different isoforms and a cathepsin L-like protease were predominantly identified proteinases. Subcellular localization of identified proteins was predicted to be either extracellular or cytoplasmic, with smaller numbers localized/co-localized in the nucleus, plasma membrane and/or mitochondria. Obtained results and integration of transcriptomic and proteomic datasets provided comprehensive snapshot for future studies aimed at potential roles of different molecules and for establishing novel strategies for the treatment and control of *F. magna* infections (Cantacessi et al. 2012).

References

Adlard RD, Barker SC, Blair D, Cribb TH (1993) Comparison of the second internal transcribed spacer (ribosomal DNA) from populations and species of Fasciolidae (Digenea). Int J Parasitol 23:423–425. doi:10.1016/0020-7519(93)90022-Q

Andrés JA, Bogdanowicz SM (2011) Isolating microsatellite loci: looking back, looking ahead. In: Orgogozo V, Rockman MV (eds) Molecular methods for evolutionary genetics. Humana Press, New York

Andrews RH, Chilton NB (1999) Multilocus enzyme electrophoresis: a valuable technique for providing answers to problems in parasite systematics. Int J Parasitol 29:213–253. doi:10.1016/S0020-7519(98)00168-4

Ansorge WJ (2009) Next-generation DNA sequencing techniques. N Biotechnol 25:193–205. doi:10.1016/j.nbt.2008.12.009

Avise JC (1994) Molecular markers, natural history and evolution, 2nd edn. Chapman & Hall, New York

Avise JC, Walker D (1999) Species realities and numbers in sexual vertebrates: perspectives from an asexually transmitted genome. Proc Nat Acad Sci USA 96:992–995. doi:10.1073/pnas.96.3.992

Balloux F, Lugon-Moulin N (2002) The estimation of population differentiation with microsatellite markers. Mol Ecol 11:155–165. doi:10.1046/j.0962-1083.2001.01436.x

Bazsalovicsová E, Králová-Hromadová I, Špakulová M, Reblánová M, Oberhauserová K (2010) Determination of ribosomal internal transcribed spacer 2 (ITS2) interspecific markers in *Fasciola hepatica*, *Fascioloides magna*, *Dicrocoelium dendriticum* and *Paramphistomum cervi* (Trematoda), parasites of wild and domestic ruminants. Helminthologia 47:76–82. doi:10.2478/s11687-010-0011-1

Bazsalovicsová E, Králová-Hromadová I, Radvánszky J, Beck R (2013) The origin of the giant liver fluke, *Fascioloides magna* (Trematoda: Fasciolidae) from Croatia determined by high-resolution melting screening of mitochondrial *cox*1 haplotypes. Parasitol Res 112:2661–2666. doi:10.1007/s00436-013-3433-0

Bazsalovicsová E, Králová-Hromadová I, Štefka J, Minárik G, Bokorová S, Pybus M (2015) Genetic interrelationships of North American populations of giant liver fluke *Fascioloides magna*. Parasit Vectors 8:288. doi:10.1186/s13071-015-0895-1

Bennett CD, Campbell MN, Cook CJ, Eyre DJ, Nay LM, Nielsen DR, Rasmussen RP, Bernard PS (2003) The LightTyper: high throughput genotyping using fluorescent melting curve analysis. BioTechniques 34:1288–1295

Bildfell RJ, Whipps CM, Gillin CM, Kent ML (2007) DNA-based identification of a hepatic trematode in an elk calf. J Wildl Dis 43:762–769. doi:10.7589/0090-3558-43.4.762

Birky CW Jr (2001) The inheritance of genes in mitochondria and chloroplasts: laws, mechanisms, and models. Ann Rev Genet 35:125–148. doi:10.1146/annurev.genet.35.102401.090231

Bombarová M, Marec F, Nguyen P, Špakulová M (2007) Divergent location of ribosomal genes in chromosomes of fish thornyheaded worms, *Pomphorhynchus laevis* and *Pomphorhynchus tereticollis* (Acanthocephala). Genetica 131:141–149. doi:10.1007/s10709-006-9124-3

Bombarová M, Vítková M, Špakulová M, Koubková B (2009) Telomere analysis of platyhelminths and acanthocephalans by FISH and Southern hybridization. Genome 52:897–903. doi:10.1139/g09-063

Bombarová M, Špakulová M, Koubková B (2014) New data on the karyotype and chromosomal rDNA location in *Paradiplozoon megan* (Monogenea, Diplozoidae), gill parasite of chubs. Parasitol Res 113:4111–4116. doi:10.1007/s00436-014-4082-7

Boore JL (1999) Animal mitochondrial genomes. Nucleid Acid Res 27:1767–1780. doi:10.1093/nar/27.8.1767

Burger G, Gray MW, Lang BF (2003) Mitochondrial genomes: anything goes. Trends in Genet 19:709–716. doi:10.1016/j.tig.2003.10.012

Buschiazzo E, Gemmell NJ (2006) The rise, fall and renaissance of microsatellites in eukaryotic genomes. BioEssays 28:1040–1050. doi:10.1002/bies.20470

Cantacessi C, Mulvenna J, Young ND, Kašný M, Horák P, Aziz A, Hofmann A, Loukas A, Gasser RB (2012) A deep exploration of the transcriptome and excretory/secretory proteome of adult *Fascioloides magna*. Mol Cell Proteomics 11:1340–1353. doi:10.1074/mcp.M112.019844

Collins FH, Paskewitz SM (1996) A review of the use of ribosomal DNA (rDNA) to differentiate among cryptic *Anopheles* species. Insect Mol Biol 5:1–9

Crimi M, Rigolio R (2008) The mitochondrial genome, a growing interest inside an organelle. Int J Nanomed 3:51–57. doi:10.2147/ijn.s2482

Dieringer D, Schlötterer C (2003) Two distinct modes of microsatellite mutation processes: evidence from the complete genomic sequences of nine species. Genome Res 13:2242–2251. doi:10.1101/gr.1416703

Dobigny G, Ducroz JF, Robinson TJ, Volobouev V (2004) Cytogenetics and cladistics. Syst Biol 53:470–484. doi:10.1080/10635150490445698

Etter PD, Bassham S, Hohenlohe PA, Johnson EA, Cresko WA (2011) SNP discovery and genotyping for evolutionary genetics using RAD sequencing. Methods Mol Biol 772:157–178. doi:10.1007/978-1-61779-228-1_9

Feagin JE (2000) Mitochondrial genome diversity in parasites. Int J Parasitol 30:371–390. doi:10.1016/S0020-7519(99)00190-3

Flegr J (2009) Evoluční biologie, 2nd edn. Academia, Praha (in Czech)

Gibbons JG, Branco AT, Yu S, Lemos B (2014) Ribosomal DNA copy number is coupled with gene expression variation and mitochondrial abundance in humans. Nat Commun 5:4850. doi:10.1038/ncomms5850

Guichoux E, Lagache L, Wagner S, Chaumeil P, Léger P, Lepais O, Lepoittevin C, Malausa T, Revardel E, Salin F, Petit RJ (2011) Current trends in microsatellite genotyping. Mol Ecol Res 11:591–611. doi:10.1111/j.1755-0998.2011.03014.x

Gundry CN, Vandersteen JG, Reed GH, Pryor RJ, Chen J, Wittwer CT (2003) Amplicon melting analysis with labelled primers: a closed tube method for differentiating homozygotes and heterozygotes. Clin Chem 49:396–406. doi:10.1373/49.3.396

Halton DW (2004) Microscopy and the helminth parasite. Micron 35:361–390. doi:10.1016/j.micron.2003.12.001

Hartl DL, Freifelder D, Snyder LA (1988) Basic genetics. Jones and Bartlett Publishers, Boston

He R, Kim MJ, Nelson W, Balbuena TS, Kim R, Kramer R, Crow JA, May GD, Thelen JJ, Soderlund CA, Gang DR (2012) Next-generation sequencing-based transcriptomic and proteomic analysis of the common reed, *Phragmites australis* (Poaceae), reveals genes involved in invasiveness and rhizome specificity. Am J Bot 99:232–247. doi:10.3732/ajb.1100429

Hearne CM, Ghosh S, Todd JA (1992) Microsatellites for linkage analysis of genetic traits. Trends Genet 8:288–294. doi:10.1016/0168-9525(92)90256-4

Hillis DM, Davis SK (1986) Evolution of ribosomal DNA: fifty million years of recorded history in the frog genus *Rana*. Evolution 40:1275–1288. doi:10.2307/2408953

Hillis DM, Dixon MT (1991) Ribosomal DNA: molecular evolution and phylogenetic inference. Q Rev Biol 66:411–453

Hörweg C, Prosl H, Wille-Piazzai W, Joachim A, Sattmann H (2011) Prevalence of *Fascioloides magna* in *Galba truncatula* in the Danube backwater area east of Vienna, Austria. Wien Tierärztl Mschr 98:261–267

Hu M, Chilton NB, Gasser RB (2004) The mitochondrial genomics of parasitic nematodes of socio-economic importance: recent progress and implications for population genetics and systematics. Adv Parasitol 56:133–212. doi:10.1016/S0065-308X(03)56003-1

Josko D (2012) Updates in immunoassays: parasitology. Clin Lab Sci 25:185–190

Karamon J, Larska M, Jasik A, Sell B (2015) First report of the giant liver fluke (*Fascioloides magna*) infection in farmed fallow deer (*Dama dama*) in Poland—pathomorphological changes and molecular identification. Bull Vet Inst Pulawy 59:339–344. doi:10.1515/bvip-2015-0050

Kobayashi T (2011) Regulation of ribosomal RNA gene copy number and its role in modulating genome integrity and evolutionary adaptability in yeast. Cell Mol Life Sci 68:1395–1403. doi:10.1007/s00018-010-0613-2

Králová-Hromadová I, Špakulová M, Horáčková E, Turčeková L, Novobilský A, Beck R, Koudela B, Marinculić A, Rajský D, Pybus M (2008) Sequence analysis of ribosomal and mitochondrial genes of the giant liver fluke *Fascioloides magna* (Trematoda: Fasciolidae): intraspecific variation and differentiation from *Fasciola hepatica*. J Parasitol 94:58–67. doi:10.1645/GE-1324.1

Králová-Hromadová I, Bazsalovicsová E, Štefka J, Špakulová M, Vávrová S, Szemes T, Tkach V, Trudgett A, Pybus M (2011) Multiple origins of European populations of the giant liver fluke *Fascioloides magna* (Trematoda: Fasciolidae), a liver parasite of ruminants. Int J Parasitol 41:373–383. doi:10.1016/j.ijpara.2010.10.010

Králová-Hromadová I, Bazsalovicsová E, Demiaszkiewicz AW (2015) Molecular characterization of *Fascioloides magna* (Trematoda: Fasciolidae) from south-western Poland based on mitochondrial markers. Acta Parasitol 60:544–547. doi:10.1515/ap-2015-0077

Le TH, Blair D, McManus DP (2000) Mitochondrial genomes of human helminths and their use as markers in population genetics and phylogeny. Acta Tropi 77:243–256. doi:10.1016/S0001-706X(00)00157-1

Le TH, Blair D, McManus DP (2001) Complete DNA sequence and gene organization of the mitochondrial genome of the liver fluke, *Fasciola hepatica* L. (Platyhelminthes; Trematoda). Parasitology 123:609–621

Lotfy WM, Brant SV, DeJong RJ, Le TH, Demiaszkiewicz A, Rajapakse RP, Perera VB, Laursen JR, Loker ES (2008) Evolutionary origins, diversification, and biogeography of liver flukes (Digenea, Fasciolidae). Am J Trop Med Hygeine 79:248–255

Lydeard C, Mulvey M, Aho JM (1989) Genetic variability among natural populations of the liver fluke *Fascioloides magna* in white-tailed deer, *Odocoileus virginianus*. Can J Zool 67:2021–2025. doi:10.1139/z89-287

Minárik G, Bazsalovicsová E, Zvijáková Ľ, Štefka J, Pálková L, Králová-Hromadová I (2014) Development and characterization of multiplex panels of polymorphic microsatellite loci in giant liver fluke *Fascioloides magna* (Trematoda: Fasciolidae), using next generation sequencing. Mol Biochem Parasit 195:30–33. doi:10.1016/j.molbiopara.2014.06.003

Morgan EA (1982) Ribosomal RNA genes in *Escherischia coli*. In: Busch H, Rothblum L (eds) The cell nucleus: rDNA. Academic Press, New York

Mulvey M, Aho JM, Lydeard C, Leberg PL, Smith MH (1991) Comparative population genetic structure of a parasite (*Fascioloides magna*) and its definitive host. Evolution 45:1628–1640. doi:10.2307/2409784

Nadler SA, De Leon GPP (2011) Integrating molecular and morphological approaches for characterizing parasite cryptic species: implications for parasitology. Parasitology 138:1688–1709. doi:10.1017/S003118201000168X

Nadler SA, Lindquist RL, Near TJ (1995) Genetic structure of Midwestern *Ascaris suum* populations: a comparison of isoenzyme and RAPD markers. J Parasitol 81:385–394

Naem S, Budke CM, Craig TM (2012) Morphological characterization of adult *Fascioloides magna* (Trematoda: Fasciolidae): first SEM report. Parasitol Res 2:971–978. doi:10.1007/s00436-011-2582-2

Nei M, Rooney AP (2005) Concerted and birth-and-death evolution of multigene families. Ann Rev Genet 39:121–152. doi:10.1146/annurev.genet.39.073003.112240

Nguyen P, Sahara K, Yoshido A, Marec F (2010) Evolutionary dynamics of rDNA clusters on chromosomes of moths and butterflies (Lepidoptera). Genetica 138:343–354. doi:10.1007/s10709-009-9424-5

Noller HF (1991) Ribosomal RNA and translation. Ann Rev Biochem 60:191–227. doi:10.1146/annurev.bi.60.070191.001203

Novobilský A, Kašný M, Mikeš L, Kovařčík K, Koudela B (2007) Humoral immune responses during experimental infection with *Fascioloides magna* and *Fasciola hepatica* in goats. Parasitol Res 101:357–364. doi:10.1007/s00436-007-0463-5

Nunome T, Negoro S, Miyatake K, Hirotaka Y, Fukuoka H (2006) A protocol for construction of microsatellite enriched genomic library. Plant Mol Biol Rep 24:305–312. doi:10.1007/BF02913457

Oberhauserová K, Bazsalovicsová E, Králová-Hromadová I, Major P, Reblánová M (2010) Molecular discrimination of eggs of cervid trematodes using the Teflon (PTFE) technique for eggshell disruption. Helminthologia 47:147–151. doi:10.2478/s11687-010-0022-y

Okimoto R, Macfarlane JL, Clary DO, Wolstenholme DR (1992) The mitochondrial genomes of two nematodes, *Caenorhabditis elegans* and *Ascaris suum*. Genetics 130:471–498

Prokopowich CD, Gregory TR, Crease TJ, Gregory TR, Crease TJ, Crease TJ (2003) The correlation between rDNA copy number and genome size in eukaryotes. Genome 46:48–50. doi:10.1139/g02-103

Pyziel AM, Demiaszkiewicz AW, Kuligowska I (2014) Molecular identification of *Fascioloides magna* (Bassi, 1875) from red deer from South-Western Poland (Lower Silesian wilderness) on the basis of internal transcribed spacer 2 (ITS-2). Pol J Vet Sci 17:523–525. doi:10.2478/pjvs-2014-0077

Queller DC, Strassmann JE, Hughes CR (1993) Microsatellites and kinship. Trends Ecol Evol 8:285–288. doi:10.1016/0169-5347(93)90256-O

Radvánský J, Bazsalovicsová E, Králová-Hromadová I, Minárik G, Kádaši L (2011) Development of high-resolution melting (HRM) analysis for population studies of *Fascioloides magna* (Trematoda: Fasciolidae), the giant liver fluke of ruminants. Parasitol Res 108:201–209. doi:10.1007/s00436-010-2057-x

Reblánová M, Špakulová M, Orosová M, Bazsalovicsová E, Rajský D (2010) A description of karyotype of the giant liver fluke *Fascioloides magna* (Trematoda, Platyhelminthes) and review of Fasciolidae cytogenetics. Helminthologia 47:69–75. doi:10.2478/s11687-010-0012-0

Reblánová M, Špakulová M, Orosová M, Králová-Hromadová I, Bazsalovicsová E, Rajský D (2011) A comparative study of karyotypes and chromosomal location of rDNA genes in important liver flukes *Fasciola hepatica* and *Fascioloides magna* (Trematoda: Fasciolidae). Parasitol Res 109:1021–1028. doi:10.1007/s00436-011-2339-y

Ririe KM, Rasmussen RP, Wittwer CT (1997) Product differentiation by analysis of DNA melting curves during the polymerase chain reaction. Anal Biochem 245:154–160. doi:10.1006/abio.1996.9916

Sattmann H, Hörweg C, Gaub L, Feix AS, Haider M, Walochnik J, Rabitsch W, Prosl H (2014) Wherefrom and whereabouts of an alien: the American liver fluke *Fascioloides magna* in Austria: an overview. Wiener Klinische Wochenschrift 126:23–31. doi:10.1007/s00508-014-0499-3

Schlötterer C (2004) The evolution of molecular markers—just a matter of fashion. Nat Rev Genet 5:63–69. doi:10.1038/nrg1249

Simon C, Pääbo S, Kocher TD, Wilson AC (1990) Evolution of mitochondrial ribosomal RNA in insects as shown by the polymerase chain reaction. In: Cleeg M, O'Brien S (eds) Molecular evolution. UCLA Symposia on molecular and cellular Biology, new series. Liss, New York

Tighe PJ, Ryder RR, Todd I, Fairclough LC (2015) ELISA in the multiplex era: potentials and pitfalls. Proteomics Clin Appl 9:406–422. doi:10.1002/prca.201400130

Wittwer CT, Reed GH, Gundry CN, Vandersteen JG, Pryor RJ (2003) High-resolution genotyping by amplicon melting analysis using LCGreen. Clin Chem 49:853–860. doi:10.1373/49.6.853

Wolstenholme DR (1992) Animal mitochondrial DNA: structure and evolution. Int Rev Cytol 141:173–216

Yang Y, Ma H (2009) Western blotting and ELISA techniques. Researcher 1:67–86

Zalapa JE, Cuevas H, Zhu H, Steffan S, Senalik D, Zeldin E, McCown B, Harbut R, Simon P (2012) Using next-generation sequencing approaches to isolate simple sequence repeat (SSR) loci in the plant sciences. Am J Bot 99:193–208. doi:10.3732/ajb.1100394

Zane L, Bargelloni L, Patarnello T (2002) Stretegies for microsatellite isolation: a review. Mol Ecol 11:1–16. doi:10.1046/j.0962-1083.2001.01418.x

Zheng X, Chang QC, Zhang Y, Tian SQ, Lou Y, Duan H, Guo DH, Wang CR, Xing-Quan Zhu XQ (2014) Characterization of the complete nuclear ribosomal DNA sequences of *Paramphistomum cervi.* Sci World J Article ID 751907. doi:10.1155/2014/751907

Final Conclusions and Future Perspectives

Since its first discovery in 1875, *Fascioloides magna* has been intensively studied throughout the last 140 years. Different periods of investigation were characterized by different types of studies. The end of the 19th and the very beginning of the 20th centuries were focused on taxonomic classification, systematic revisions and determinations of various scientific names of giant liver fluke, until Henry B. Ward proposed the new genus *Fascioloides* with the type species *Fascioloides magna* in 1917. Detailed morphology and description of all stages of the life cycle of *F. magna* were provided in the first half of the 20th century, with significant contribution of William E. Swales from Canada and Božena Erhardová-Kotrlá from the Czech Republic.

The 70s, 80s and 90s of the last century represented a very active period focused mainly on two types of studies; pharmacological tests of an efficacy of different anthelmintic drugs, and experimental infections of various North American intermediate and final hosts. The important contribution in this scientific field has to be addressed to William J. Foreyt from Washington, USA. In Europe, the parasitologists from the Czech Republic have significantly contributed to the knowledge on spectrum of intermediate hosts of giant liver fluke.

A great effort has been made to determine *F. magna* in free-living and domestic ruminants throughout the North American enzootic regions. Its occurrence has been monitored since the early 1930s. In Europe, first discoveries of *F. magna* started in Italy in 1875, followed by findings of the parasite in the Czech Republic in 1910 and in the Danube region in the 1990s. Spatial distribution of the parasite in Europe is evidently dynamic and ongoing process. The latest findings of *F. magna*, in particular in two distinct localities in Poland (Lower Silesian Wilderness and Podkarpackie Province), and in northeastern Bavaria in Germany, are alarming. The fascioloidosis is spreading beyond permanent natural foci and this process is likely to continue. Therefore, studies on the host range of parasite, changes in their distribution and regular monitoring in already established foci and in potentially high-risk neighbouring areas, are reasonable and their continuity has to be maintained in order to prevent the infection of farmed and domestic ruminants.

A reasonable question to ask would be: "How can we prevent a spread of *F. magna* infection?". To bring forward such question is easier than to find an appropriate

I. Králová-Hromadová et al., *The Giant Liver Fluke, Fascioloides magna: Past, Present and Future Research*, SpringerBriefs in Animal Sciences,
DOI 10.1007/978-3-319-29508-4

answer to it. Theoretically, the prevention measures can be focused on both intermediate and final hosts. Aquatic snails, intermediate hosts of *F. magna*, usually occupy fragile water biotopes. Intervention into these biotopes (e.g. drainage), sometimes proposed as a possible prevention measure, would not provide an ecology-friendly solution. Natural movement of cervids is impossible to stop and moreover, their treatment can not be well controlled and directed. The only possible measurement can be addressed to man-made translocation of cervids and their veterinary and parasitological control. However, in vivo diagnostics, based on faecal examination and determination of *F. magna* eggs in stool, is meaningful only in definitive hosts. In order to eliminate the spread of infection in domestic ruminants, a possible precaution is to avoid sharing pastures with confirmed fascioloidosis.

With rocket development of molecular techniques (e.g. the whole genome sequencing), the modern approaches were implemented in studies of *F. magna*, providing very useful outcomes. Although the first sequence of ribosomal ITS2 spacer was acquired in 1993, the molecular "boom" has been recorded since 2007 and carries on to date. Molecular markers, such as selected regions of ribosomal and mitochondrial genes, were proved to be useful in molecular taxonomy and identification of the species and in determination of genetic interrelationships among and between North American and European populations. The design of microsatellite markers has opened up new possibilities for molecular genetics, more detailed population structuring, and determination of transmission and migratory routes, what will certainly be applied in a near future.

The future perspective lies undoubtedly within a deeper insight of parasite's genome/transcriptome/proteome studies. Important data on transcriptome analysis were already furnished and the next challenge is to determine the roles of particular molecules, mainly those important for treatment and control of giant liver fluke.

The past and present research of *Fascioloides magna* brought many important and interesting data on biology, distribution, host-parasite interrelationships, therapy, immunology and molecular biology. We believe that future research will produce some novel and remarkable information that will contribute to better understanding of this fascinating parasite with "never-ending story".